WORKS ISSUED BY

The Hakluyt Society

THE VOYAGES OF CADAMOSTO

SECOND SERIES

No. LXXX

ISSUED FOR 1937

COUNCIL
OF
THE HAKLUYT SOCIETY
1937

Sir William Foster, C.I.E., *President*.
The Right Hon. The Earl Baldwin of Bewdley, K.G., *Vice-President*.
Admiral Sir William Goodenough, G.C.B., M.V.O., *Vice-President*.
James A. Williamson, Esq., D.Lit., *Vice-President*.
E. W. Bovill, Esq.
Sir Richard Burn, C.S.I.
G. R. Crone, Esq.
Vice-Admiral Sir Percy Douglas, K.C.B., C.M.G.
E. W. Gilbert, Esq., B.Litt.
Vincent T. Harlow, Esq., D.Litt.
A. R. Hinks, Esq., C.B.E., F.R.S.
T. A. Joyce, Esq., O.B.E.
Malcolm Letts, Esq.
Prof. A. P. Newton, D.Lit.
N. M. Penzer, Esq.
Prof. Edgar Prestage, D.Litt.
S. T. Sheppard, Esq.
Brigadier-General Sir Percy Sykes, K.C.I.E., C.B., C.M.G.
Roland V. Vernon, Esq., C.B.
R. A. Wilson, Esq.
Edward Heawood, Esq., *Treasurer*.
Edward Lynam, Esq., M.R.I.A., *Hon. Secretary* (British Museum, W.C.).
The President
The Treasurer } *Trustees*.
William Lutley Sclater

S. JORGE DA MINA. FROM A CHART BY SEBASTIÃO LOPES, 1558
(*British Museum, Add. MS. 27,303*)

THE
VOYAGES OF CADAMOSTO

AND OTHER DOCUMENTS ON
WESTERN AFRICA IN THE SECOND HALF
OF THE FIFTEENTH CENTURY

Translated and edited by

G. R. CRONE

LONDON
PRINTED FOR THE HAKLUYT SOCIETY
MCMXXXVII

PRINTED IN GREAT BRITAIN

CONTENTS

Preface . . . . *page* ix

Introduction
   I. The trade between the Western Sudan and the Mediterranean xi
   II. The Portuguese on the coast of Western Africa, 1448–90 xvii
   III. Alvise Cadamosto xxx
   IV. The discovery of the Cape Verde Islands . xxxvi

Notes on the texts xliii

The Voyages of Alvise Cadamosto and Pero de Sintra   1

The Letter of Antoine Malfante from Tuat, 1447 . 85

The Voyages of Diogo Gomes 91

Extracts from the 'Decadas da India' of João de Barros   103

Appendix. Notes on the natural history of Cadamosto's narrative 149

Bibliography 151

Index 155

MAPS AND ILLUSTRATION

S. Jorge da Mina. (From the Chart of Sebastião Lopes, 1558. B.M. Add. MS. 27,303) . . *Frontispiece*

Western Africa and the Cape Verde Ids. from the Atlas of Grazioso Benincasa, 1468. (B.M. Add. MS. 6390) . . . . . . *opposite p.* 84

North-western Africa in the fifteenth century . . *at end*

PREFACE

THE importance of Cadamosto's narrative has long been recognized by students of African historical geography. It is clear, however, from a study of the literature, that considerable misconception is current concerning his achievements and the character of his work. This is largely due to the lack of a critical edition of the text, and of a modern translation. The sole English translation, in a condensed form, appeared in the middle of the eighteenth century.

It was thought appropriate to add to the present version translations of other documents comprising our authorities for this important period in Portuguese expansion, so that the volume might form a sequel to the edition of Azurara's 'Chronicle' by C. R. Beazley and Edgar Prestage (Hakluyt Soc., Ser. I, vols. XCV and C).

Without the assistance received from others, I should not have been able to carry through the task of editorship. In particular I am much indebted to Mr E. W. Bovill, who has co-operated throughout. He is the author of the first section of the Introduction, and of a number of footnotes. Indeed, it is at his wish alone that his name does not appear on the title-page as co-editor. I am also grateful to Professor Edgar Prestage for his generous advice and help; and to Senhor Armando Cortesão for considerable assistance in translating the extracts from Barros. As all editors of volumes for the Hakluyt Society will appreciate, Mr Edward Lynam, the Honorary Secretary, has rendered me many kindnesses throughout.

G. R. CRONE

June 1937

INTRODUCTION

I

The trade between the Western Sudan and the Mediterranean

THE course of the Portuguese exploration of the western coast of Africa has been narrated by Sir Raymond Beazley and Professor Prestage in their introduction to the Hakluyt Society edition of Azurara's *Chronicle of the Discovery and Conquest of Guinea*[1], although that work ends approximately at the year 1448. The documents printed in the present volume carry on the narrative from that date until the closing years of the century, and in this introduction emphasis will be laid upon the relations between these coastal voyages and the trade of the interior, which had been highly organized on well-defined routes for a considerable period before the beginning of Portuguese expansion.

Throughout the Middle Ages there had been a thriving trade between the Christians of Europe and the Moslems of the Maghreb, as they called North Africa. Christian galleys were constantly putting in and out of a dozen or more African ports, of which Massa, Saffi, Salee and Tangier on the Atlantic, and Honein, Algiers, Bone and Tunis on the Mediterranean, were the most important. The Normans of Sicily had been amongst the first in the field, but later the Pisans, Genoese, Marseillais and Venetians acquired their respective rights and privileges along the Barbary coast. Each nation had negotiated treaties to protect its nationals and its interests, and maintained a consul and a *fonduk* in the ports with which it traded. So firmly was this trade established that it was seldom interrupted by the scourge of piracy—a constant source of international friction, in which Christians, more often than Moslems, seem to have been the aggressors.

[1] Hakluyt Soc., First series, vols. xcv (1896) and c (1899).

In spite of their consuls and their *fonduks*, the Christians were rigorously excluded from the interior. Occasionally an enterprising merchant might penetrate as far as Marrakech, Constantine or Kairwan, but here his treaty rights gave him no protection against the middlemen who jealously guarded their control over the trade of the interior against the interference of interlopers. Consequently there was a marked contrast between the extraordinary journeys of Europeans in Asia and their complete ignorance of the interior of northern Africa, although they had long known it to be a region of great commercial activity.

That there was great wealth in the remote interior was patent enough, for the articles which attracted Christian merchants to the Barbary ports were not products of the coastal belt. The most important of these were gold, negro slaves, malaguetta pepper or chillies, which so puzzled Europeans that they called it grains of Paradise, ivory, ebony, and gaily dyed goat skins. For these the Christians eagerly bartered their glass beads, cloth and the miscellaneous trade goods which probably differed but little from those in demand in the remoter parts of Central Africa to-day. Themselves confined to the coast, the Christians found in the Jews a valuable source of information about the trade of the interior. The Jews had long played an important part in the commercial life of northern Africa, and from their *mellahs* on the coast had spread into the oases of the Sahara and thence into the Sudan.

In the fourteenth century the Jewish cartographers of Majorca had produced several remarkable maps of Africa containing information about the interior, which till then had never been made known. The source from which they got it was evidently their co-religionists in Africa. The most notable of these maps was the so-called Catalan Atlas made for Charles V by Abraham Cresques in 1375, on which appeared such names as *tenbuch* (Timbuktu), *ciutat de melli* (Mali), *geugeu* (Gao) and *tagaza*, which were soon to become famous. The Atlas Mountains are shown broken by a pass used by 'merchants going to the Land of the Negroes of Guinea'. Across the centre of the Sahara appears the figure of a negro monarch enthroned with a sceptre in one hand and a nugget of gold in the other. 'This negro lord', runs the legend, 'is called Musa Mali, Lord of the Negroes of Guinea. So abundant is the gold which is found in his country that he is the richest and most noble King in all the land.'

Gold was the article of trade which most interested Jews and Christians alike and was shortly to become still more important. By the beginning of the fifteenth century the demands of a rapidly increasing foreign trade and a series of disastrous wars had greatly denuded Europe of precious metals. Consequently the period was one of financial stringency which continued till the discovery of the American mines. Meanwhile men looked for relief to Africa, whence they had long imported gold, brought overland from an unknown source beyond the Sahara. It is not surprising therefore that the fifteenth century witnessed many determined efforts by sea and land to discover whence the gold came.

Most of the gold of North Africa was shipped from the port of Massa, the ancient Temest, in the extreme west on the Atlantic coast, and in the fourteenth century the trade was chiefly in the hands of the Pisans and Venetians. Massa owed its importance to its proximity to Marrakech, the northern terminus of the great gold route of the Sahara. From Marrakech the road crossed the Atlas, as shown on the Catalan map, to Sijilmasa, which, lying on the edge of the desert, had long enjoyed importance as an entrepôt for trade and a fitting out place for caravans. As early as the tenth century it was known for its importance in the gold trade. Ibn Haukal, Masudi, Yakut and Idrisi all bear testimony to this. From Sijilmasa the road ran south to the salt mines of Taghaza[1], some twenty days' march away.

For many centuries these mines played an important part in the economic and political life of the interior of northern Africa. Although in many respects so richly endowed by nature, the negro countries of the south were always hard put to it to obtain adequate supplies of salt. Shut off from the sea by dense forests, and with only a few hopelessly inadequate salt deposits of their

[1] This Taghaza, in 25° N., is not to be confused with two other sources of salt of the same name, but of much less importance. Midway between this Taghaza and Taudeni was Taghaza el Ghizhan or Taghaza of the Gazelles. Further west, in 23° N., are deposits to-day called Sebka d'Ijil formerly known as Taghaza el Gharbie, or West Taghaza.

own, they were dependent for this necessity of life on what they could import. This circumstance has been exploited to the full by foreign merchants throughout history down to the present day. For this reason the control of the Taghaza mines was a matter of the first importance to Morocco.

El Bekri in the eleventh century says Taghaza was constantly thronged with merchants, and that the mines were producing an enormous revenue. Three centuries later Ibn Battuta gave the following description of what he saw there:

After twenty-five days, we reached Taghaza, an unattractive village, with the curious feature that its houses and mosques are built of blocks of salt, roofed with camel skins. There are no trees there, nothing but sand. In the sand is a salt mine; they dig for the salt, and find it in thick slabs....No one lives at Taghaza except the slaves of the Massufa tribe, who dig for the salt; they subsist on dates imported from Dar'a and Sijilmasa, camels' flesh, and millet imported from the Negrolands. The negroes come up from their country and take away the salt from there. At Iwalatan (Walata) a load of salt brings eight to ten *mitkal*, in the town of Malli it sells for twenty to thirty, and sometimes as much as forty. The negroes use salt as a medium of exchange, just as gold and silver is used (elsewhere); they cut it up into pieces and buy and sell with it. The business done at Taghaza, for all its meanness, amounts to an enormous figure in terms of hundredweights of gold-dust[1].

An anonymous writer of the twelfth century describes the bartering of salt for gold as follows:

In the sands of that country is gold, treasure inexpressible. They have much gold, and merchants trade with salt for it, taking the salt on camels from the salt mines. They start from a town called Sijilmasa... and travel in the desert as it were upon the sea, having guides to pilot them by the stars or rocks in the deserts. They take provisions for six months, and when they reach Ghana they weigh their salt and sell it against a certain unit of weight of gold, and sometimes against double or more of the gold unit, according to the market and the supply[2].

From Taghaza the road continued south across the desert, probably dividing and offering two alternative routes, one

[1] Ibn Battuta, *Travels in Asia and Africa*, ed. H. A. R. Gibb, Lond., 1929, p. 317.
[2] *Tohfut ul Alabi*, in Palmer, H. R., *Sudanese Memoirs*, II, p. 90.

through Walata, a market of Soninke origin which enjoyed a considerable period of prosperity, and the other farther to the east leading to Timbuktu.

Timbuktu first became an important market as early as the eleventh century. The notable part it so long played in the commercial life of the interior of north-western Africa was due to its geographical position. Situated close to the navigable waterway of the Niger and on the threshold of the desert, it was the meeting place of those who travelled by water with those who travelled by land—the people of the Sudan and the people of the desert. The former brought gold, grain and kolanuts which they exchanged for the salt, dates, and merchandise of the Maghreb. By the end of the thirteenth century it had become an important entrepôt for the trade between Jenne, higher up the Niger, and Walata, and was trading not only with all parts of the Maghreb but also with Egypt. The road through Timbuktu followed the Niger up to Jenne, the Soninke city long noted as a centre of trade and culture. Surrounded by a network of waterways, Jenne had long preserved independence of the powerful neighbouring state of Mali, the capital of which—Niani—was only a few days' march to the south-west.

Mali or Melli was a Mandingo kingdom which had attained considerable importance in the previous century under Mansa Musa, or Kankan Musa, who extended his kingdom eastwards to include the greater part of the Middle Niger. Mansa Musa acquired a European reputation as the result of a spectacular pilgrimage to Mecca in 1324, when he dazzled Cairo with his prodigal display of wealth. As we have already seen, the impression he then made earned him a reputation sufficient to win him a prominent place and a tribute to his wealth on the Catalan map of Abraham Cresques who calls him Musa Mali, and he continued to appear on the principal world maps until the sixteenth century.

His wealth, like that of the independent city of Jenne, was due to the proximity of Wangara. To-day this name survives only as that of a Moslem branch of the Mandingo people, but for centuries it was the name of the great gold-bearing districts of Bambuk and Bure, bounded on the north by the Senegal, on the west by the Faleme, on the east by the Niger and on the south by the Tinkisso.[1] It was this region to which the great trans-Sahara gold route led, either by way of Timbuktu or Walata. It was for centuries the goal of all who travelled this ancient road, and somewhere here was the scene of the silent bartering of salt for gold. This curious traffic was first described by Herodotus; in the tenth century Masudi found that they knew of it in Sijilmasa; Yaqut described it two centuries later, and finally we have Cadamosto's account. Although this curious method of silent barter is chiefly associated in men's minds with the West African gold trade, it has been recorded in many other parts of the world. It appears usually to have arisen through difficulties of language, or perhaps more often when dire necessity drove a timid people to trade with dreaded and more powerful neighbours. In Wangara salt was the supreme need of the people, and perhaps for nothing else would they have been brought to trade their gold. The story that they would exchange gold for an equal weight of salt is not as improbable as it sounds.

Wangara was probably not the only source from which Barbary drew the gold which so attracted European merchants. Possibly at this period gold was being worked at Lobi on the Black Volta, where there are gold workings of great age. South of Lobi too were the Ashanti gold fields, which may also, but less probably, have been contributing their quota.

Nor was the Taghaza road the only route by which gold

[1] The gold of the negroes was obtained from sedimentary deposits of pliocene and pleistocene age widely distributed throughout western Africa, and from river gravels washed out of older formations. In addition to the above area, extensive primitive workings are found in the Bouré and Sieké valleys in Upper Guinea, in the Lobi district, and throughout the Gold Coast. In the sedimentary deposits, shafts were sunk through the laterite cover to a depth of three or four times the height of a man, the women carrying away and washing the 'dirt'. The work was done between January and May, to avoid flooding of the pits during the rainy season. Nuggets of considerable size were sometimes found; one weighing 30 kilograms was found at Diébélé at the beginning of this century. In the rivers, the gold was obtained by diving during the low-water season. (A description of native methods will be found in Meniaud (*Haut-Sénégal-Niger*, II, pp. 166 ff.). The use of the pits no doubt gave rise to the legend that the gold was produced by gigantic ants, and the percolation of fresh sediment into disused pits was the basis of the negro belief that gold 'grew'.

reached the Mediterranean. There were other caravan routes, each one doubtless playing its part in the gold trade. One road appears to have run from Timbuktu to Tuat and Wargla, and thence to Tuggurt and through the El Kantara gorge to Constantine. Farther east there was another road running south from Tunis through Ghadames and Ghat to Gao on the Niger.

Gao, Cadamosto's Cochia and the capital of Songhai, was destined shortly to succeed to the predominant position in north-western Africa long held by Mali. The original Songhai capital was Kukia, probably situated in Dendi to the east of the Niger and close to the present north-west frontier of Nigeria. Early in the eleventh century the capital was transferred to Gao, some four hundred miles farther north on the left bank of the Niger. It seems probable, however, that Kukia remained the name by which the Songhai capital was known to foreigners. It was used by Idrisi in the twelfth century, and in the middle of the fifteenth century not only by Cadamosto but by one of his successors on the coast, Diego Gomes, who referred to a great city in the interior called Quioquun, which can scarcely have been other than the Cochia of Cadamosto. In later years Gao became immensely wealthy by reason of its stores of gold, and it is therefore not improbable that in the fifteenth century, when it was already of commercial as well as political importance, it was consigning gold to the north, drawing its supplies from Lobi, and perhaps from other gold fields farther west.

II

The Portuguese on the coast of Western Africa, 1448–90

Prince Henry's resolve to devote his energies to the exploration of Africa probably dated from 1415 when, aged twenty-one years, he won his spurs at the capture of Ceuta. In Africa he first heard of the ancient caravan traffic of the Sahara bringing gold, slaves, ivory, and ebony from the remote countries of the negroes, already known as Guinea. It was this rich trade which kept the ports of Barbary thronged with Christian galleys bartering the trade goods of Europe with the Moorish merchants who controlled this traffic. Though at various times during the fifteenth century the directors of Portuguese policy toyed with the idea of territorial expansion in northern Africa, with the object of securing the trans-Saharan traffic for themselves, an alternative method and one promising more success, was to attempt to establish contact with the sources of this wealth by sea, and so divert trade from the land routes and the Moorish middlemen[1].

It is probable that some knowledge of the coast of Africa as far as the Gulf of Guinea was current in western Europe at that time. The Catalan Atlas of 1375, which may be taken as a typical fourteenth-century cartographical document, displays some slight acquaintance with the trade routes and markets of the Niger basin. More significantly it records a voyage by Jaume Ferrer in 1346 along the coast in search of the semi-legendary Rio de Oro. From another contemporary document, the *Libro del Conoscimiento* of the anonymous Spanish Franciscan, it is possible to obtain a glimpse of the trading activities of the Moors along this coast, and to deduce that these extended as far as the Gulf of Guinea[2]. It is therefore very probable that the delineation of the great gulf on the coast of West Africa, which first appears in the world maps of Sanuto at the beginning of the fourteenth century and reappears in maps one hundred years later, before the Portuguese had pushed thus far to the south, was based upon a substratum of fact.

The rediscovered island groups of Madeira and the Canaries added to the awakening interest in this quarter and formed suitable bases for voyages along the western coast, though the

[1] This motive is, in fact, attributed to Prince Henry by Dr J. Munzer, who moved in official Portuguese circles, and may here be recording a tradition. 'Knowing that the King of Tunis, that is, of Carthage, obtained much gold each year, he (Prince Henry) sent spies to Tunis, and having ascertained that this king despatched merchants to southern Ethiopia who exchanged their goods for slaves and gold, determined to do by sea what the king of Tunis had done for many years by land.' Munzer, *Itinerario*, ed. B. de Vasconcellos, p. 141.

[2] See Taylor, E. G. R., 'Pactolus, river of gold' (*Scottish Geogr. Mag.* XLIV (1928), p. 129), and Cortesão, J., 'O designio do Infante' (*Hist. do Portugal*, vol. III).

Portuguese never succeeded in establishing themselves firmly in the latter.

The voyages initiated by Prince Henry were not, therefore, thrusts into the unknown, but part of a sustained attempt to wrest control of an important economic artery then in alien and often hostile hands. This detracts in no way from their achievement, for its execution demanded courage, skill, and determination of the highest order.

The control of this trade was no doubt Prince Henry's initial objective. During the same period, interest in the mysterious Christian priest-king, Prester John, was increasing. His identity had puzzled Europe for centuries, but he was at this period believed to be the Negus of Abyssinia. As the Portuguese came to learn of the various potentates of north central Africa, they were eager, on very slender grounds, to identify them in turn with this monarch. The problem was not finally solved until the journey of Covilhã, begun in 1487. Since Prester John was regarded as a ruler of 'the Indies', the attempts to establish contact with him no doubt widened the goal of the explorations, and thus led to the circumnavigation of Africa and the opening up of the route to India at the close of the century[1].

The narratives printed in this volume continue the history of Portuguese enterprise on the western coasts of Africa from the point to which it was carried by Azurara in his *Chronicle of the Discovery and Conquest of Guinea* to the closing years of the century. Azurara's narrative deals with events which took place approximately before 1448. The extent of coast-line explored by the Portuguese by that year is difficult to determine with accuracy. In Santarem's opinion their navigators had already reached Sierra Leone: 'on voit...que les navigateurs portugais poussèrent les explorations avant 1448 à cent dix lieues au delà du cap Vert, et conséquemment au delà de la Sierra Leoa, qu'ils reconnurent le Rio Grande, le Rio do Lago, et d'autres points[2].'

[1] An ambassador from Prester John is said to have been in Lisbon in 1452; 'Jorge embaçador do preste Joham que lhe mandamos pera mantyamente de hũ mes' (quoted from Peres, D., 'O caminho da India' (*Hist. do Portugal*, III, p. 565). If so, it is strange that the Portuguese were not better informed as to the situation of his country.

[2] Santarem, Vcte de, *Recherches sur la priorité, etc.*, p. 294. J. Cortesão considers that Cape Palmas had been reached before the death of Prince Henry.

This statement is based upon Azurara's narrative; he relates that Gil Eannes had pushed on beyond Cape Verde for sixty leagues, reaching a point where 'all the land seemed like marshes'. This description might well be applied to the islands and low-lying land around the estuary of the river later known as the Rio Grande. He also records that some years later Alvaro Fernandes had voyaged one hundred and ten leagues beyond Cape Verde, his farthest point being 'a narrow strip of sand stretching in front of a great bay', where he went in by boat to the beach. If the distance given is correct, this would not have brought him as far as Sierra Leone. It is difficult to believe, also, that no note would have been made by Azurara of the discovery of Sierra Leone by Fernandes or another contemporary navigator, for these mountains, in contrast to the general low level of the coast, greatly impressed the early navigators. Another remark of Azurara's, that from Fernandes' farthest point the coast-line 'trendeth commonly to the South', does not suggest that Sierra Leone had been reached, or at least, passed for any considerable distance, for it is at Cape Saint Anne that the coast-line takes a decided south-easterly direction.

The names given to coastal features by these two navigators have not survived. Azurara records no names after Cape Verde, with the exception of the Cabo dos Mastos, a few miles only beyond the Cape, and a Rio de Nuno, the scene of Nuno Tristão's death in 1446, which was probably the river later called the Rio Grande, the modern Jeba.

It appears fair to say, therefore, that by 1448 the Portuguese were approaching Sierra Leone and had begun the detailed examination of the coast-line between Cape Verde and the latter landmark. It is important to realise this, so that the voyages of the next decade can be appreciated in their true light, that is, as the completion of the work of their predecessors, and as commercial ventures, rather than as voyages of discovery. There seems, therefore, to have been an undoubted pause in the progress of Portuguese exploration during this decade, for there is no evidence of a definite advance until the voyage of Pedro de Sintra, beyond Sierra Leone, generally assigned to 1462, and the completion of the exploration of the Cape Verde

Islands, probably in 1459, by Antonio da Noli, and their colonisation. Several reasons may be adduced for this. Diogo Gomes says that the work of exploration was interrupted for two years by the expedition to Alcacer el Seguer in 1458. In the years 1451-4 Portugal was also engaged in war with Castile over the Canaries. The difficulties of Prince Henry in regard to the financing of these expeditions and his death in 1460 may also be reflected in this period of consolidation rather than of advance. The stimulus to renewed activity was undoubtedly the granting in 1469 by King Affonso of the licence to Fernão Gomes.

The first document for what may be called the post-Azurara period is the narrative of the Venetian, Alvise da Ca' da Mosto, or Cadamosto, as he is known to English writers. Cadamosto sailed from Cape Saint Vincent on board a caravel belonging to Vincente Dias on March 22nd, 1455[1], and after a quick run made the island of Porto Santo. After a brief stay here and at Madeira, they continued to the Canary Islands, visiting Gomera and Ferro. Cadamosto's description of the Canaries is one of the earliest, being preceded only by those of Azurara and the clerics who accompanied Bethancourt's expedition, with which he was not likely to have been acquainted when he wrote. Though he notes their commercial possibilities, his interests were wider than those of a mere trader. He recorded geographical and historical details, and was particularly intrigued by the lives and customs of their inhabitants, the doomed Guanches.

From the Canaries, the voyage was continued first due south, to avoid the difficulties of the African coast, and then southeasterly to sight Cape Blanco. Cadamosto does not state that he

---

[1] There has in the past been some doubt as to the date of Cadamosto's first voyage. Several writers, for example Mendes Trigozo, the translator of the Portuguese version of 1812, place it in 1444. This appears to rest upon the authority of Damião de Goes. The date given in the text, however, is supported by other evidence, e.g. the data collected by A. da Mosto (see below, p. xxxi), the date of the sailing of the Venetian galleys, and the statements, for what they are worth, in Usodimare's letter (see below, p. xxiii, note). Snr. J. Cortesão, however, suggests that if the voyages are assigned to 1444 and 1445, this would eliminate the discrepancy between the number of caravels taking part in the discovery before 1446 as given by Azurara, and the number otherwise known (*O sigilo*, p. 76).

visited Arguim, though at this point he inserts a description of the island and its trade with the interior[1].

Arguim, close to the African mainland, was a Portuguese base, and had been fortified some time after 1445. Prince Henry had been granted a monopoly of the African trade, and the merchants were trading here under his licence. Here they were in contact, through Arab merchants and the desert tribes, with the market of Wadan, a stage on one of the caravan routes between Morocco and the Sudan. In return for gold dust and negro slaves, the nomads obtained cotton, cloth, and other trade goods, and especially corn, for their food supplies were always scanty. Cadamosto describes in some detail the Azanaghi, the Azaneguys of Azurara, known to-day as the Sanhaja, one of the most important of the veiled tribes of Tuareg who still inhabit the central and western Sahara, and, probably from Moorish sources, the markets and trade of the interior. The story of the traffic in gold naturally excited his interest most, and he is the first European, with the possible exception of Malfante, to put on record a comprehensive account of its character and ramifications, including the famous 'silent traffic' with the negroes of the south. Something has already been said of this traffic; Cadamosto is a witness that the Portuguese attempt to divert it from Morocco to the coast was meeting with a certain degree of success.

From Cape Blanco he voyaged southwards to the mouth of the Senegal. Like the Portuguese mariners in whose tracks he was following, he was greatly struck by the contrast between the regions north and south of the river, between the arid lands, the home of brown-skinned nomads, on the north, and the bush and forest of the south, peopled by negroes. His narrative conveys vividly the first impressions of a European on encountering an environment differing so completely from those of western and Mediterranean Europe.

He notes that in the Senegal region Muhammadanism, through the agency of traders, had won a certain number of converts, principally among the ruling classes, but was not firmly established.

---

[1] Nordenskiöld erroneously asserts that Cadamosto made 'a peaceful expedition of discovery to the interior' from Arguim (*Periplus*, p. 120).

His first sojourn on the African coast was made in the territory of a chief called Budomel, to the north of Cape Verde, probably the ruler of Cayor. This chief was already well known to the Portuguese for his integrity, and under his patronage Cadamosto found a ready market for the horses he had brought with him, the demand being attributed by him to the high mortality due to the great heat.

After leaving Budomel, Cadamosto continued towards Cape Verde, and fell in on the way with two caravels, one of which was commanded by a Genoese, Antoniotto Usodimare[1]. They ultimately reached the estuary of the Gambia, and attempted to sail upstream, but the hostility of the negroes dashed their hopes of obtaining gold, and they returned to Portugal.

The following year Cadamosto, again in company with Usodimare, and with a third caravel equipped by Prince Henry, sailed for the Gambia. On the voyage they sighted and examined cursorily the Cape Verde Islands. On arriving in the Gambia they were successful in establishing friendly relations with a negro chief, Battimansa, tributary to the Emperor of Mali. Some bartering resulted, but the yield of gold was disappointing. It is obvious that these negroes, having nothing of value to offer in exchange, were obtaining little gold from 'Wangara'. A small quantity of ivory was forthcoming. An outbreak of fever among the crews brought the visit to the Gambia to an end. Before returning to Portugal, however, the caravels sailed as far south as the Rio Grande, but, as the interpreters were unable to make themselves understood, no trade resulted from this venture. The width of the estuary of the Rio Grande

[1] Antoniotto Usodimare, a member of an ancient family prominent in the service of Genoa from the twelfth century, was born in 1416. After holding office in his native city, he went as a merchant first to Seville and afterwards to Portugal. He is the author of the curious letter (first brought to light by Gräberg de Hemso), in which he describes a voyage to the Guinea coast in 1455, that is, the voyage during which he encountered Cadamosto, and states that he was preparing for a second venture. His letter therefore contributes to the dating of Cadamosto's voyages, though he does not mention the latter by name. The letter, which also relates his alleged encounter with a descendant of a member of the Vivaldi expedition, is printed and critically discussed by Magnaghi, in *Precursori di Colombo*? He died sometime before 1462. He has sometimes been assumed to be identical with A. da Noli, but this is erroneous (Caddeo, R., *Navigazioni Atlantiche*, p. 87).

impressed Cadamosto considerably, and the emphasis he placed upon it in his narrative is reflected in its delineation on Benincasa's charts.

On these two voyages, with the exception of his sighting of the Cape Verde Islands, Cadamosto made no new discoveries. The value of his work lies elsewhere. It is the first original account to have survived of a voyage into the regions opened up by European enterprise at the dawn of modern overseas expansion, and reflects the spirit of open-minded enquiry characteristic of the new age. The fabulous and the sensational have no place in the story he has to tell. His outlook was singularly comprehensive, and he was evidently at pains to collect and co-ordinate information from many sources. That his work remained for a considerable period one of the primary authorities on western Africa is testimony to its thoroughness.

The narrative of Diogo Gomes[1], the next voyager to be considered, has come down in an unsatisfactory state, which is accounted for by the fact that it was taken down by Martin Behaim from an oral statement by Gomes many years after the events with which it deals, and Behaim's version was later translated into Latin by Valentim Fernandes. The narrative terminates about the year 1475 and was apparently taken down by Behaim in 1482. It is not, therefore, to be wondered at that it displays considerable confusion. The opening portion covers the period dealt with by Azurara, and is therefore not reprinted here. He relates two voyages he made to the Guinea coast, on the second of which he claimed to have discovered the Cape Verde Islands. It is difficult to establish with any certainty the date of the first voyage. He says that it took place 'not a long time after' the disastrous expedition of Vallarte the Dane. The date of this was 1448. Later, he says that for two years after his return no one went to Guinea as the Prince was occupied with the expedition to Alcacer el Seguer, which took place in 1458. The voyage therefore took place either in 1457, or, at the

---

[1] Gomes was born *circa* 1420. After serving as a page to Prince Henry, he was knighted in 1440. In 1466 he was a magistrate at Cintra, and sixteen years later was warden of the Castle there. (Ravenstein, *Martin Behaim* p. 32.)

earliest, in 1456, which, it should be noted, was the date of Cadamosto's second voyage[1].

The caravels, according to the Behaim-Fernandes narrative, proceeded first to the mouth of the Fancaso (the present Rio Grande). After some trading with the natives—in which they secured a small quantity of pepper—the other captains prevailed on Gomes to turn back as they were alarmed by the strength of the sea currents. The next landfall they made was apparently the Cabo dos Mastos. After going ashore for a short while, they resumed a southerly course, and on the next day made the estuary of the Gambia. In the following days Gomes made friends with the people of Frangazik, the nephew of Farisangul, on the right bank of the river. He was informed that the negroes on the left bank would not make friends with the Portuguese because they had slain Nuno Tristão and his companions. Gomes' recollection seems to be at fault here for Nuno was slain on the Rio Grande, and it has been suggested that this is a reference to the hostility encountered by Cadamosto on his first voyage. Gomes in one caravel, taking with him a negro, Bucker, proceeded up the river a considerable distance to Cantor, beyond which it was impossible to proceed farther owing to the narrowing of the river. At Cantor he was told of the trade routes to Timbuktu and Kukia. He also learnt of the watershed dividing the westward-flowing coastal rivers from those flowing to the east. This information was of considerable importance, for it tended to dispose of the theory that the Senegal system was connected with that of the 'Western Nile' or Niger. This divide was formed by the 'Serra Geley' or Mount Gelu, on the opposite side of the range called 'Serra Lyoa'. From Mount Gelu it was said that gold was obtained. In this region there was also a great river Emin, and a 'lake' which appears to have been from his description an area inundated by a river. This was probably the Niger. He heard here of the wars between two chiefs, Sambegeny and Semanagu, news of which he says had

[1] Senhor A. Cortesão places this voyage in 1456. For his views upon the relations of Cadamosto's and Gomes' voyages, see Cortesão, A. Z., *Subsidios para a história do descobrimento da Guiné e Cabo Verde*, to which further reference is made later.

reached Prince Henry from a merchant of Oran. The lord of the region was 'Bormelli', resident at Kukia. Regarding the method of procuring the gold, he was told that men dug it from pits, the sand then being washed for gold by women. Owing to the impure air, presumably in the pits, the men did not live long.

On returning down the river to the other caravels, he found that fourteen men had died from the heat, and many were sick: he therefore continued to the river mouth. Here he made friends with the chief Battimansa, alleged to have been hostile to the Christians. Battimansa, however, shifted the blame on to Nomimansa, the lord of the land about the estuary. The latter proved as eager as Battimansa to make friends, and went so far as to beg Gomes to baptise him, a request which was diplomatically refused. Gomes, however, promised that the Prince would send out a priest and establish firm relations with him. Gomes then left the Gambia on his return voyage to Portugal. When off Cape Verde, he encountered two canoes, in one of which was Beseghichi, lord of that land. Gomes says that he took the opportunity to reprove him for his treatment of Christians. It was with this chief that Pero de Evora concluded peace in 1481. Then continuing on his voyage, Gomes put in at Arguim and finally reached Lagos.

Gomes' second voyage, he states, took place 'two years afterwards', but the event to which this phrase refers is not clearly stated. If his landfall at the Cape Verde Islands is to be accepted, it must have taken place before 1460, when the islands are first referred to (see below). He mentions the pause in the course of discovery occasioned by the expedition to Alcacer (El Qsar es Sgir). This town was taken in 1458, so that his second voyage must have been made between that year and 1460[1].

He first touched at the port of Zaya in the land of the Barbacins, i.e. to the north of the Gambia. Here he fell in with the caravels of Gonzalo Ferreira and Antonio da Noli, a Genoese, who were trading in horses. Owing to the growth of this trade, the value of a horse in terms of slaves had fallen considerably. Gomes then received news that a certain de Prado, apparently a Spanish interloper, was returning from the Gambia with a

[1] A. Cortesão (*op. cit.*) puts Gomes' second voyage in the year 1460.

richly laden caravel, the fruits of trading arms with the Moors. In virtue of the authority invested in him by the King before his departure, Gomes despatched Ferreira to arrest de Prado. This Ferreira accomplished, and on the return to Portugal de Prado was executed.

Gomes, in company with da Noli, then sailed from Zaya homewards. After two days and a night they sighted islands in the sea. On one of these Gomes was the first to land, and to it he gave the name of Santiago. These were the Cape Verde Islands. After passing the Canary Islands and Madeira, Gomes was driven by a contrary wind to the Azores; da Noli, more fortunate, made a quicker passage to Portugal, forestalled Gomes with the news of the discovery and received the Captaincy of the Island of Santiago.

There is little to be said of Pedro de Sintra's voyage of 1462 beyond the fact that in the course of it he coasted from the Rio Grande past Sierra Leone to Cape Cortese, as is recorded in Cadamosto's version of his log. De Sintra's claim to be the discoverer of Sierra Leone is confirmed by Duarte Pacheco's incidental remark on the significance of this name[1]. The style of the narrative as preserved by Cadamosto closely follows that of the earliest surviving Portuguese *roteiros*, from the beginning of the sixteenth century. This supports the contention that the Portuguese had begun as early as the middle of the fifteenth century to compile these sailing directions, and it may be that this version of de Sintra's voyage should be regarded as a survival of a primitive pilot[2].

It is curious that, with the exception of da Noli, Barros mentions none of these navigators. The earlier portion of his history, as he explicitly states, is merely a rewriting of Azurara's chronicle. His contribution really begins with a statement of da Noli's discovery of the Cape Verde Islands, the date of which he fixes in or about 1461[3]. The completion of the exploration of the archipelago was accomplished by Diogo Affonso

[1] See below, p. 81, note.
[2] See Fontoura da Costa, A., *A marinharia dos descobrimentos*, p. 278.
[3] Galvão says that this voyage took place in 1462. His brief entry corresponds closely with the details recorded by Barros (Galvão, *Discoveries of the World* (Hakluyt Soc. vol. xxx), p. 73).

in the following year. The question of the discovery of these islands is discussed in detail later.

The next forward step in the Portuguese advance was initiated when the King in 1469 leased the monopoly of trade on the west African coast to Fernão Gomes for five years, on condition that he secured the exploration of one hundred leagues of new coastline beyond Sierra Leone each year. Gomes paid for this privilege two hundred thousand *Reis* a year. Later he purchased the trading rights at Arguim for one hundred thousand *Reis* a year. The King reserved to himself the option to purchase all the ivory at a fixed price. The immediate result of this contract was the voyage of João de Santarem and Pero de Escobar in 1471 along the Gold Coast to the point near which the fortress of El Mina was afterwards built, followed in 1474 by that of Sequeira to Cape Saint Catherine, and of Fernão do Po to the island now known by his name. When Gomes' lease expired in 1474, it was not renewed because, it is suggested, he was not in a position to defend these waters from Spanish inroads.

From 1475 to 1480, Portugal was engaged in wars with Castile, and this explains the slackening of the West African enterprise after the expiration of Gomes' lease. During these years, numerous Spanish vessels appeared off the Guinea coast, seeking to win a share in the trade. Diogo Gomes refers to the activities of one as early as 1459, and in 1483 Diego Cão captured three Spaniards in West African waters. On the accession of John II, and the conclusion of the wars with Spain, renewed attention was paid to the African trade. Partly to facilitate this, and in part to check the activities of the interlopers, it was decided in 1481 to build a fortress, known later as 'S. Jorge da Mina' ('The Mine' of English writers), on the Gold Coast. The site chosen was favourably placed in relation to the auriferous deposits of the Ankober and neighbouring river valleys. This was the first gold-producing area with which the Portuguese had established direct contact. The fortress of Arguim had failed to divert effectively the traffic in gold between the 'Wangara' and the upper Lobi area and North Africa. The opening up of the Gambia, though leading them closer to the former area, had not brought the Portuguese the wealth they sought, as the traffic still remained in the hands of intermediaries. Barros' account of the organisation of the slave trade is important. Not merely were the Portuguese carrying them off to Europe, but they were also soon operating as middlemen in this traffic. Their ships bore the captives from the source of supply in Benin and the Congo to Fernão do Po, whence they were despatched to the market at El Mina to be sold to the Mandingo traders from the north. Thus all pretence that the trade was justified by the spiritual salvation of the captives had to be dropped. The new energy behind the direction of African affairs was also shown by the despatch of Diogo Cão's expedition which reached the mouth of the Congo in 1483. The exploration of the west coast of Africa was now in its last stages.

After describing the building of the fort of El Mina by Diogo de Azambuja, Barros relates circumstantially the misfortunes of the Jalof Prince, Bemoy. This account is interesting for the light it throws upon the spread of Portuguese influence in western Africa. Though the subsidiary project, the building of a fort at the mouth of the Senegal, failed, Barros tells us that tales of the King of Portugal's power disposed the native rulers to greater respect for his agents. However this may be, Barros records in the following years Portuguese missions to Timbuktu, Tucurol (Tekrur), the kingdom of Mali in Mandinga, and to Wadan. The latter failed to achieve its aim, the establishment of a factory, and the fact that such a move was considered necessary emphasises the failure of Arguim as a centre of the gold traffic.

By the close of the century, therefore, the Portuguese had established contact with the rulers of western Africa, and so drawn into their hands a considerable portion of the traffic in gold which had previously found its way to Europe through Moorish intermediaries. In addition to this commodity, they were also receiving ivory, pepper, civet and other exotic products, and, above all, slaves. In return they sent out, besides the usual trumperies that have always attracted the unsophisticated, small quantities of European manufactures, such as cloth and silk, and the horses so much in demand in the hostile environment of the tropics. The sensational results of the opening of the sea route to India and the East, and the subsequent overwhelming demands upon their resources, soon led to a decline in their power in western Africa, and the other nations of Europe were quick to assail it. But it was many years before knowledge of the interior progressed appreciably beyond the point to which it had been brought by the Portuguese pioneers.

III

Alvise da Ca' da Mosto

Until Signor A. da Mosto published the results of his researches in the archives of Venice, very little was known of Cadamosto's career beyond the statements in his narrative[1]. Although these researches have added nothing to our knowledge of the voyages, they have value in confirming some of his statements, particularly the date of his departure from Italy, and thus, by implication at least, the date of his first voyage. They possess a greater value, however, in so far as they permit a more complete appreciation of the character and achievements of the man, apart from his African voyages, which enables a judgement on his general trustworthiness to be reached, a judgement, it may be said at once, entirely favourable to his reputation. There is also the satisfaction to be derived in gleaning a little more of the life of one whose fame has been firmly established by the achievements of scarcely more than two years.

The origin of his family has not been ascertained with certainty, but it appears probable that he was descended from a branch of the da Mosto family of Lodi which, during the troubles of the twelfth century, had fled to Venice. His father, Giovanni da Ca' da Mosto, married Giovanna Querini in 1428, and by her had four sons and two daughters. The date of the birth of Alvise is uncertain; he himself implies that he was born in 1432. In the proof required for admission to the *Maggior Consiglio* in 1451, his age was given as 25, i.e. he was born in 1426. Signor da Mosto is inclined to dismiss this, partly

---

[1] Mosto, A. da, 'Il navigatore Alvise da Mosto' (*Archivio Veneto*, II (1927), pp. 168-259). I have used the conventional form of the name, as it is now well established.

because it conflicts with the date of his parents' marriage, and partly because it was based merely on the testimony of witnesses. Apart from this contention, it will be seen later that Alvise was engaged in business from 1442 onwards, and made a voyage in 1445. This, though not conclusive, points to a birth date nearer 1426 than 1432. The elder Cadamosto in his earlier years seems to have been a citizen of some standing. He acquitted himself creditably at the siege of Verona by the Milanese in 1440, and three years later was appointed a 'Savio agli Ordini', an administrator of the war and mercantile marine. He had, however, a weakness for litigation, and this ultimately brought about his downfall.

After a few years as a business associate of Andrea Barbarigo, Alvise sailed in 1445 on one of the latter's Barbary galleys, and the following year made the voyage to Candia. Five years later he was elected a 'nobile balestriere' on the great galleys of Alexandria. Each of these galleys carried six noble bowmen, who were quartered in the poop and messed at the Commander's table, and twenty-six common bowmen. This appointment suggests that he was not by training a navigator, and that his later interest in overseas voyaging was commercial, a conclusion which is supported by the few facts known of his career. Senhor Bensaude has suggested that Cadamosto was later commissioned by King Affonso to take observations in connection with the compilation of navigational tables, but there is no record to substantiate this. In 1452, Alvise sailed with the Flanders galleys in the same capacity. On his return to Venice the next year, he found his family in distress. His father, attempting to revenge himself for a failure in a civil suit which had brought upon him a fine and interdiction from the public service, had become involved in a complicated intrigue, the upshot of which was a further fine, and banishment from the territories of Venice. These were the circumstances in which Alvise determined to embark upon another voyage, and Signor da Mosto is justified in supposing that they also weighed with him in his subsequent decision to seek a career in Portugal. Accordingly, he embarked with his younger brother Antonio—the 'relative' of his narrative—upon one of the three Flanders galleys of

1454, under Marco Zeno. The departure date which he gives, August 8th, is not quite accurate; the three galleys sailed on the 11th, 12th and 13th of August respectively; the 8th was the date of the promulgation of the Senate's order. After his West African voyages, the events of which have been outlined above, Alvise left Portugal for Venice in February 1463/4; his brother Antonio, who had rejoined him in 1460, remained abroad until 1465 as Venetian consul at Seville. On his return, Alvise, with his father's consent, assumed the administration of the family estates, engaged in commerce, and in the service of the State. He married in 1466 Elisabetta Venier, with whom he received a dowry of 2000 ducats. In the following years it is interesting to find him engaged in a suit over a quantity of 'orchel' or orchella weed, and exporting malmsey to London. After serving in several minor offices—he was elected an 'Uditor novo', a civil appeal judge, in 1470—he was called to responsible offices which he filled until his death. Venice, engaged in a protracted and unsuccessful struggle with the Turks, had sought allies among the princes of the Balkans. It was to one of these, Duke Vlatco, of Santa Sava, ruler of Herzegovina, that in 1474 Cadamosto was sent as ambassador extraordinary. Later in the same year, he assisted in putting Cattaro into a state of defence to resist a threatened Turkish attack. Again, in 1476 he was called upon to serve abroad, this time as Governor of Corone, one of the key fortresses of Venetian power in the Morea. There he served for three years, taking his share in the negotiations which brought the war with Turkey to a close. His later offices included the command of the trading galleys sent to Alexandria in 1481. His death occurred while on an official mission to Rovigo in July 1483.

The date of the composition of Cadamosto's narrative is to some extent doubtful. From the statement in the introductory chapter, if this is his work, it would appear that he was writing from memory, and not putting together notes written during the voyages. This view is adopted by Professor Almagià, who considers that the narrative was written after Cadamosto's return to Italy, i.e. after 1463. This has been employed to explain certain discrepancies as to dates and distances in the text, about

INTRODUCTION xxxiii

which his memory might have been inaccurate, but these can be explained more plausibly as the errors of the copyist, for the autograph manuscript has not survived. Certain considerations, also, support the view that the whole text, as contained in the *Marciana* manuscript and printed in the *Paesi*, is unlikely to have been composed from memory. The account of de Sintra's voyage resembles closely the style of a book of sailing directions, perhaps compiled from the explorer's log, and does not suggest an account written from memory. It is probable that Cadamosto brought back to Venice some such document from which he extracted his account. Moreover, some report of his voyages had reached Venice before his return, as appears from the cartographical evidence. The doge, Foscarini, who died in 1457, is said to have expressed the hope that Cadamosto's success and the map of Fra Mauro would influence the King of Portugal to persevere in his project. Fra Mauro's map, though it does not appear to show cartographically any relation with Cadamosto's voyages, has two legends which seem almost certainly to be inspired by his narrative. The Cape Verde Islands are not shown, nor is there any trace of the nomenclature of de Sintra's voyage. If the map, as is generally assumed, was completed in 1460, the latter omission is explained.

It appears therefore that portions at least of the narrative had been composed before 1460, that is, before Cadamosto left Portugal. There he would be able to collect information from the Portuguese, for example, about the Canary Islands, for his account can scarcely be the result of a single visit. The cartographical evidence points to the narrative, with the addition of de Sintra's voyage, having been put in its present form in, or shortly before, 1468.

The evidence for a close relationship between Cadamosto's narrative, including that of de Sintra's voyage, and the work of a contemporary cartographer, Grazioso Benincasa, is strong. The chart by Andrea Bianco dated 1448 is assumed to have been prepared to illustrate the progress of Portuguese discovery. The complete chart has been published by Fisher, and a photograph of the portion relevant to this discussion was reproduced in the *Geographical Journal* for 1895. The extreme limit of the

West African coast shown is Cape Verde and Cape Roxo, with a bay between, beyond which the coast is shown conventionally as running due east. Thus the map shows approximately the results up to the year of its compilation. The next cartographical evidence of importance is the great world map by Fra Mauro in the Biblioteca Marciana at Venice. Material for this was provided by King Afonso, and he also paid for the execution of a copy, which was sent to him through Stepan Trevisan in 1459. It should also be noted that Andrea Bianco assisted in its compilation.

On this map the west coast of Africa terminates at the great gulf running eastwards which is characteristic of cartography from at least the beginning of the fourteenth century. The most southerly point bears the inscription 'Nota che dal cavo verde in suso non se vede la tramontana'. To the north appear the names 'c⁰ rosso', 'c⁰ verde', 'cauo del bori', 'c⁰ palmear', and 'cauo de vertude'. A great river, representing the combined Senegal-Niger, enters the ocean by two mouths north of Cape Verde; to the south is an unnamed river, which may represent the Gambia, and beyond a smaller river which may be Cadamosto's 'Rio Grande'. As far as the littoral is concerned, therefore, the Fra Mauro map displays little advance upon Bianco's chart of 1448.

The note about the North Star mentioned above may have been inspired by a passage in Cadamosto's narrative, but there is another legend on the Fra Mauro map which can almost certainly be assigned to this source. It occurs in the centre of the Sahara, and reads 'Qui fra terra sono alguni negri che hano i labri grandissimi permodo che li convien portar sopra queli sal achoche non se putrefaça e questi son queli che baratano oro per sal. el suo consueto e de vignir a certo tempo a uno hiogo (? luogo) deputado a questo barato e qui metono al incontro del sal tanto oro quanto li par...', and continues to describe this silent traffic. The mention of the negroes with large lips and their 'putrefaction' suggests the use of Cadamosto's account.

It is in the later work of Grazioso Benincasa that an extensive use of Cadamosto's material is found. Down to 1468 Benincasa's atlases show no appreciable advance upon Bianco's chart of 1448. A chart, however, dated in that year, makes extensive use of Cadamosto's, and particularly of de Sintra's, nomenclature, and the Cape Verde Islands are shown for the first time on any map. A comparison of the names recorded by Cadamosto with those occurring on this chart of Benincasa's shows that the latter must have had the Venetian traveller's account before him. On Cadamosto's first voyage, he tells us that it was he who gave the name 'Rio de Barbacini' to the modern Joal; the name 'Barbacis', apparently applied to a river, occurs twice on Benincasa's chart, and 'bodumel' is also to be found there. The phrase found on this chart, 'Casa de li Rey', is probably inspired by Cadamosto's 'Casa sua', though it has been transposed south of Cape Verde. Of the four names recorded in the second voyage, and which are claimed to have been then bestowed for the first time, namely 'Isola de Sancto Andrea', 'Rio de Casamansa', 'Rio de Sancta Anna', and 'Rio de Sancto Dominico', two—'Rio de Casamansa' and 'Rio de san domingo'—appear on the Benincasa chart. In addition the latter has a large number of names not employed by Cadamosto, some of which may be the relics of earlier Portuguese exploration.

The correspondence between the de Sintra narrative and the Benincasa chart is more striking. The fifteen names on the coast recorded in the former from the 'Rio de Besegue' to the 'Fiume de le Palme', have their counterparts on the chart. The outline of the coast also corresponds closely with its description given in the narrative, so that it is clear whence Benincasa drew his information. Cadamosto asserts that he himself had drawn a chart of his voyages: this has not survived, and it is open to doubt whether he had the technical ability to produce a chart. If, however, this statement is correct, the chart must have corresponded closely to Benincasa's[1]. A small point is worth noting as showing the close connection between Cadamosto's narrative and Benincasa's atlases. There are two atlases of his dated 1468. On one of them the 'Arbori di Sancta Maria' has been added in the extreme south. In Cadamosto's narrative, too, the reference to this wood is inserted at the end, out of its proper place. It almost appears as though the reference was added to the original manuscript and the name inserted on the

[1] The names on Benincasa's chart are in Portuguese.

map by one and the same person[1]. One may conclude therefore that the narrative was written between 1463 and 1468, and in all probability in the latter year. Cadamosto has been credited with the compilation of an anonymous *portolano*, that is, written sailing directions for the Mediterranean and the coasts of western Europe, but not covering the coasts of western Africa dealt with in his narrative. This work—a unique copy of the first printed edition, Venice 1490, is in the *Marciana*—is attributed to Cadamosto on the authority of Sansovino—'scrisse un libro intitolato Portolano: ma senza il suo nome'. This portolano is thoroughly discussed by Sig. A. da Mosto[2].

IV

The discovery of the Cape Verde Islands

Cadamosto's claim to have discovered the Cape Verde Islands has been challenged by many historians[3]. The arguments against accepting his narrative as accurate are stated fully by Major[4], who drew largely upon the work of Lopes de Lima[5]. Major's main points are (1) that if Cadamosto sailed from Lagos at the beginning of May, he could not have discovered the islands on May 1st; (2) that a ship blown west-north-westwards from the

---

[1] See Emiliani, M., 'Carte nautiche dei Benincasa' (*Bol. Soc. Geogr. Ital.* ser. vii, 1 (1936), p. 485).

[2] Mosto, A. da, 'Il portolano, etc.' (*Bol. Soc. Geogr. Ital.* ser. iii, vi (1893), p. 540).

[3] Barros, writing in the sixteenth century, believed that the Cape Verde Islands were the 'Insulae Fortunatae' of classical authors, but these were more probably the Canary Islands. On the strength of the two islands, labelled 'dos hermanes' on Bianco's chart of 1448, some writers have held that the Cape Verde Islands had been sighted before that date. It is clear, however, that these islands merely represent a perpetuation of the legend of the Pillars of Hercules on the Island of Gades (see note by C. E. Nowell in *Geogr. Journ.* LXXXIX (1937), p. 485). A recent writer draws attention to the existence of what may possibly be megalithic dolmens on the islands, and also to rock inscriptions, which have been tentatively identified as Berber (Chevalier, A., *Iles du Cap Vert*). The exact character of the 'dolmens' is however doubtful, and the inscriptions, if they are Berber, cannot be older than the end of the fifteenth century, and may be the work of slaves. There is no evidence therefore for a discovery of the islands before 1450.

[4] Major, R. H., *Prince Henry the Navigator*, pp. 286–7.

[5] Lopes de Lima, J. J., *Ensaios, etc.*, vol. I, pt. 2, pp. 3–7.

neighbourhood of Cape Blanco would be driven away from the islands; (3) that Santiago is not visible from Boavista; and (4) that there is no river of fresh water in any of the islands, nor are turtles or salt to be found there. Major concluded that Diogo Gomes is entitled to be regarded as their discoverer. These arguments have been repeated, with minor additions, by Portuguese historians. In support of Cadamosto, most of them have been answered by H. Yule Oldham[1], and cannot be accepted as destroying Cadamosto's claim.

With regard to the first, the oldest manuscript extant (see p. 63) gives March as the month of departure from Lagos. Though this demolishes Major's argument, however, it does not explain why he took so long to reach the islands, especially as he says that he spent no time in the Canary Islands. The second point can also be challenged, and though it has some substance, cannot be held to invalidate Cadamosto's claim. His statement is: 'Finally we reached Cape Blanco; having picked it up we stood off to sea a little. The following night there arose a storm from the south-west with a strong wind; upon which, in order not to turn back we held our course towards west-north-west, to the best of my belief, to sail as near the wind as possible, for two nights and three days. The third day we sighted land.' It is obvious of course that a vessel driven west-north-west from the neighbourhood of Cape Blanco would not sight the Cape Verde Islands. Some have attempted to defend Cadamosto by assuming a copyist's error in the direction, but this would appear also to imply another error in the direction of the wind. It seems possible, however, to accept the statement without resorting to conjectural emendation. Though the passage is very condensed, the general impression it was intended to convey is clear. When the south-west storm was encountered, the vessels, which may then have been some distance to the south of Cape Blanco, endeavoured to beat against it, instead of running before it. They succeeded to some extent in so doing, though they were carried considerably to the west. The phrase 'to the best of my belief' is noteworthy, as suggesting the doubt in Cadamosto's

---

[1] Oldham, H. Yule, *Discovery of Cape Verde Islands* (repr. fr. *Richthofen Festschrift*, 1895).

mind as to the direction of the course throughout. It probably indicates the original direction, but as the force of the storm abated, the vessels would be able to sail closer to the desired course, and it is not twisting the narrative too violently to assume that the islands were sighted when the vessels were endeavouring to regain the coast-line, which would be the first object of their pilots. Senna Barcellos rejects the story on the grounds that south-west winds are not encountered in those latitudes in May. The meteorological evidence, however, does not seem to suggest that such winds are entirely impossible at that season.

A general consideration makes it difficult to suppose that Cadamosto's account was entirely fabricated. If he were compiling a fictitious claim, he would scarcely have made an apparently elementary blunder over the position of the Cape Verde Islands relative to Cape Blanco, unless he was entirely ignorant of the true position of the island group. That he was so ignorant is extremely unlikely; on the Benincasa charts, its position is approximately correct, and as has been seen, there is little doubt that he contributed to their compilation.

The second line of criticism is directed against the accuracy of his brief description of the islands, but the evidence here is against Major's contentions.

It seems clear that he was mistaken in assuming that Santiago is not visible from Boavista. Yule Oldham quotes a passage from Professor C. Doelter describing how from the Pico d'Antonier on Santiago he saw the whole archipelago spread out before him, and this is confirmed by Senna Barcellos[1]. Major is also in error about the turtles and salt, for the presence of turtles is mentioned by many travellers, and salt became an important item in the trade of the islands. The remark about the river of fresh water may arise from a misunderstanding over the use of the term 'rio', which is sometimes used for an arm of the sea, as in 'Rio de Oro'. Cadamosto, in speaking of a river which a vessel of 150 tons could enter, and which was a bowshot wide, may therefore have been referring to a bay. There is also evidence to support Cadamosto's statement that the islands at the time of their discovery were wooded.

[1] *Subsidios para a historia de Cabo Verde*, 1900.

On the whole, it may be said that no serious fault can be found with Cadamosto's account of his discovery of the islands and his brief description of them. His narrative in general, where it can be checked independently, is trustworthy, and no reason has been suggested why he should have put forward a fraudulent claim in this particular instance. It seems, therefore, that his claim to have discovered the Cape Verde Islands in 1456 must be accepted.

As has been shown, the earliest date that can be assigned to Gomes' second voyage, if any credibility is to be attached to his narrative as recorded by Behaim and Valentim Fernandes, is 1458. It is noteworthy that Behaim, though he took the trouble to write down Gomes' narrative, made no use of it in compiling data for his globe of 1492. The translation of a legend on the globe reads 'Cape Verde Islands or Insula fortunata discovered and settled by the Portuguese in 1472'. This date is probably an error for 1462, the approximate date assigned to the discovery by Barros, and may have arisen from a confusion with the voyage of Diogo Affonso[1].

The relations between Gomes and da Noli are by no means clear. It will be remembered that Gomes states that on his return from his second voyage he was accompanied by da Noli, and that, after the landing on the islands, the latter succeeded in reaching Portugal before him and securing the credit of the discovery. It is to be observed that this account conflicts completely with the version given by Barros; in this no mention is made of Gomes, and da Noli is described as setting out from Portugal in company with his son and nephew, with three vessels. Barros also implies that the date of the discovery was *circa* 1462, and Galvão assigns it to this year.

Light is thrown upon this involved incident by official documents printed in *Alguns documentos*. The first mention is in one dated 1460, by which Prince Henry transferred the temporalities of five of the islands to the King. Two years later, in 1462, the King bestowed the twelve islands upon his brother Fer-

---

[1] Ravenstein (*loc. cit.* p. 76) points out that this mistake is also made by Waldseemüller in his world map of 1507, and suggests that both were copying a lost original.

dinand, asserting that five of them had been discovered by da Noli in the life-time of Prince Henry. In 1497, King Manuel donated part of the island of S. Thiago to D. Branca d'Aguiar, daughter of da Noli—'Mice Antonio Genovez Capitam da parte da Ribeira Grande na Ilha de S. Thiago'—stating that 'Lo dito mice Antonio foy o primeiro, que ha dita ilha achou, e comecou do povoar'[1]. Evidently, therefore, da Noli's voyage was made shortly before 1460. The date assigned above to Gomes' second voyage is 1458-9, so that a joint voyage by da Noli and Gomes at that time appears to be established. The exact share taken by each in the discovery, however, is obscure. It is difficult to understand why, if Gomes' claim was sound, he should have been passed over in favour of da Noli, at a time when the Portuguese were becoming anxious to exclude foreigners from establishing themselves in the newly discovered lands. Until further evidence is forthcoming, therefore, the effective discovery of the islands must be assigned to da Noli in 1458 or 1459[2].

This conclusion, however, does not destroy Cadamosto's claim to have made a reconnaissance of the islands three years or so previously. The interval between this visit and the expedition of da Noli[3] which resulted in the colonisation of the islands is explained by the interruption in the West African voyages caused by the expedition to Al Seguer, a circumstance to which Gomes specifically refers.

The acceptance of Cadamosto as the discoverer of the islands, a discoverer who sighted them by accident and whose achievement was followed up by others, confers perhaps no great distinction upon him. The question, had it involved this point alone, would not have merited discussion at any length. In a

[1] Torre do Tombo, *Liv. das Ilhas*, fl. 69, quoted from Lopes de Lima, *op. cit.* liv. I, pt. 2, p. 6.

[2] R. Caddeo rejects the claims of Cadamosto and Gomes in favour of da Noli.

[3] Galvão's statement about this voyage might be taken to imply that some knowledge of the existence of these islands was the basis of da Noli's request for permission to make a voyage of discovery: 'No anno de 1462 vieram a este reyno de Portugal tres Ianoeses pessoas nobres, o primeiro delles era Antā de Noly e hū seu jrmão e sobrinho, cada hū em seu nauio, pedira liceça ao Infante p'ra descobrir as ilhas do Cabo verde e elle che approuue...' (Hakluyt Soc. 1st series, xxx, p. 73).

INTRODUCTION   xli

wider view, it is, however, of importance, for historians, rejecting Cadamosto in this instance as a false claimant to the achievement of others, have used this argument to cast doubts upon his reliability in general and upon his second voyage in particular. Thus Senna Barcellos alleges that Cadamosto merely substituted his own name for that of Gomes, and Armando Cortesão believes that the narrative of his second voyage is a fabrication based upon the experiences of others. A more charitable view assumes that the attribution of the discovery to Cadamosto was the error of a careless *amanuensis*. Senhor Cortesão argues that Gomes' first voyage took place in 1456, the year in which Cadamosto claimed to have made his second voyage, and points out that both give very similar accounts of their doings on the Gambia, claiming to have been the first to establish friendly relations with Battimansa. He suggests also that when Gomes refers to the earlier hostility of the natives to Nuno Tristão, he is confusing him with Cadamosto and the brush with the natives on his first voyage. He admits that Cadamosto's description of the coast from the Casamansa southwards is accurate, but believes that it was based on Portuguese sources.

In defence of Cadamosto, it may be pointed out that in the case of de Sintra's voyage he is careful to give that navigator entire credit for his achievements. He also states correctly that before he left Portugal in 1463 others had been to the Cape Verde Islands and found that they were twelve in number—a reference to the voyage of Diogo Affonso in 1461–2. He was also aware of the names given to these new islands[1], if it is accepted that it was from him that Benincasa obtained his information. The assumption that his account of his second voyage was falsified from one by Gomes is even more untenable when it is remembered that Cadamosto's narrative was current in Venice between 1463 and 1468, while Gomes' was not written down until some years later.

A tentative solution of the problem is that Gomes took part

[1] Senna Barcellos attempts to discredit Cadamosto on the grounds that the name 'Bonavista' given by him to one of the islands did not come into general circulation until the time of King João II. Its use by Benincasa on his chart of 1468 however disposes of this argument.

in Cadamosto's second voyage—the similarity of their narratives has been mentioned already—and that his voyage to the Cape Verde Islands in 1459 or 1460 was based on the results of the earlier voyage. It is also possible that the grant in November 1457 to Prince Ferdinand of islands to be discovered in the future was prompted by Cadamosto's report. The relations of Gomes with da Noli are even more obscure, and our present knowledge admits of no final solution. These considerations, however, do not require the rejection of Cadamosto's account as untrustworthy.

To sum up—there appears no sound reason at present for challenging Cadamosto's statement that he sighted the Cape Verde Islands in 1456, while admitting that the voyage which led to the colonisation of the islands was undertaken by da Noli, perhaps in conjunction with Gomes, in 1459 or thereabouts.

NOTES ON THE TEXTS

CADAMOSTO

CADAMOSTO'S account of his two voyages, and of de Sintra's, first appeared in print in the collection, *Paesi novamente retrovati*, published at Vicenza in 1507. The introduction, which deals with the work of Prince Henry, was probably composed by the editor of the collection, for it is not found in the older manuscripts of the narrative. Two of these are extant, both now in the Biblioteca Marciana at Venice. The earlier, recently acquired by that library, has been examined and described by Professor Roberto Almagià[1]. It consists of 46 sheets in a cursive semi-gothic hand, ascribed to the second half of the fifteenth century. It is not in Cadamosto's autograph, and contains mistakes which may be attributed to the errors of a copyist. The second manuscript dates from the beginning of the sixteenth century (*Marciana. Ital.* VI. 208), and is closely related to the first printed edition.

The popularity of the *Paesi*, which contained, in addition to Cadamosto, narratives of the voyages of Columbus and Cabral, appears to have been immediate, and translations were issued in various languages, for example, a Latin edition at Milan in 1508, a German edition at Nuremberg in the same year, and a French edition at Paris in 1515. The great popularity of Cadamosto, however, is to be attributed to the Italian version published by Ramusio in the first volume of his *Navigazioni*, Venice, 1550, for it was from this source that later writers on West Africa drew. Professor Almagià does not consider that Ramusio used the text of the *Paesi*; it may be observed, however, that the differences between the two texts appear to be largely interpolations which may be the work of Ramusio. A modern editor would have printed them as footnotes. Neither Hakluyt nor Purchas included Cadamosto's voyages in their collections, pro-

[1] *Marciana. Ital.* VI. 454. Professor Almagià is preparing an edition of this text. See Almagià, R., 'Intorno ad un manoscritto dei viaggi di Alvise da Mosto' (*Revista geogr. ital.* XXXIX (1932), pp. 169–76). References in the footnotes to the *Marciana MS.* are to this codex.

bably because it was felt that the ground had been covered by John Pory's translation and amplification of Leo Africanus, undertaken at Hakluyt's instigation, and which it was originally proposed to include in the 'Principal Navigations'. Pory was of course acquainted with Cadamosto's narrative, though he did not make great use of it. Several passages in Richard Jobson's *Golden Trade* (1623) suggest that he also had read it carefully. The first English translation, somewhat abridged from Ramusio, appeared in the eighteenth century, when it was included in Thomas Astley's *New General Collection of Voyages*, vol. I, 1745. The most recent edition, with a useful introduction and notes, was published by R. Caddeo, Milan, 1928.

The translation in this volume has been made from the *Paesi* text, emended where necessary from the earlier Marciana manuscript, a photostat copy of which was courteously provided by Prof. Roberto Almagià. It has also been checked with Ramusio's version; those of his interpolations which are of sufficient interest have been inserted within square brackets. The chapter divisions and headings of the *Paesi* text have been preserved, but the text has been divided into paragraphs. The spelling of proper names in the text is that of the original, but elsewhere conventional forms have been used, and obvious misprints have been corrected.

MALFANTE

Malfante's letter was first brought to light by M. Charles de la Roncière. It forms part of a fifteenth-century collection of miscellaneous material in Latin, which includes the letters of Cicero and the journey of Nicolo dei Conti, written in two hands. This volume, now in the Bibliothèque Nationale (Nouvelles acquisitions latines, MS. 1112), consists of 140 folios, 31 × 22 cm. The letter is on folios 136 *verso*, and 137. The Latin text, with commentary and notes, was first published by M. de la Roncière in 1919 (*Comité des trav. hist., Bull. Section de Géogr.*, XXXIII, 1918 (1919), pp. 1–28), and later, with a French translation, in his *Découverte de l'Afrique*, I, pp. 143–58. The latter work reproduces a portion of the manuscript. No details of Malfante's career are known.

DIOGO GOMES

Gomes' narrative is contained in the Valentim Fernandes Codex in the Bayerische Staatsbibliothek, Munich (*Cod. Hisp.* 27). This material was collected by Fernandes for a volume of voyages which he contemplated publishing in Latin. He translated for this the German version which Martin Behaim took down from Gomes, *circa* 1482. An account of the Codex, with a transcription of the Latin text, was printed by Schmeller in 1847 (*Abhandl. phil.-philol. Klasse der K. Bayer. Ak. (München),* IV (1847)). Major's life of Prince Henry contains an English translation. In 1899, L. Cordeiro, working from a careful copy of the Munich manuscript, published a version in Portuguese (*Bol. Soc. Geogr. Lisboa,* XVII, pp. 267–93). The translation in this volume is that of Major, checked by Cordeiro's.

BARROS

The translation of the extracts from Barros has been made from the edition published at Lisbon in 1778 in nine volumes. The first two Decades were originally published at Lisbon in 1553, the third in 1563, and the fourth at Madrid in 1613. An Italian translation of the first two Decades, by A. Ulloa, appeared at Venice in 1562. The version in the present volume is given as a translation of the 1778 volume, and not as a definitive text of Barros. A critical English edition of Barros' invaluable work is still wanting.

THE VOYAGES *of* CADAMOSTO

The beginning of the book of the first navigation by the ocean to the land of the Blacks of Lower Ethiopia at the command of the *Illustrious Lord Infant Don Hurich*, brother of *Don Dourth King of Portugal*.

CHAPTER I

The first to make the navigation of the
OCEAN SEA *to the* SOUTH[1]

I, ALOUISE DA CA DA MOSTO, was the first that of the noble city of Venesia was moved to sail the ocean sea beyond the strait of Zibeltera[2] towards the south in the land of the Blacks of lower Ethiopia. On this my journey I saw many things new and worthy of some notice. In order that those that shall come after me may be able to understand what my thoughts were in the midst of varied things in strange new places,—for truely both our customs and lands, in comparison with those seen by me, might be called another world—I decided that it would be laudable to make some record of them. As my memory shall serve me, so shall I set down these matters with my pen, and if, as the facts require, they should be things not ordinarily reported, none the less they shall not want absolute truth in every respect,—as I prefer rather to understate than to relate anything which exceeds the truth. Know therefore that the first to initiate the navigation of this portion of the ocean sea towards the south of the land of the Blacks in lower Ethiopia, for from the time of our first father Adam there is no record that it was ever navigated until today (save by him of whom Pliny writes)[3],

---
[1] This introductory chapter does not appear in the two earliest manuscripts.   [2] Gibraltar.
[3] A reference to Hanno's voyage (Pliny, *Nat. Hist.*, VI, ch. 36).

was the illustrious Lord Infante Dū Hurich[1] of Portogallo son of the illustrious Don Zuāne[2] King of Portogallo. This Lord Infante Don Hurich, although many noteworthy things might be related of his virtue, I shall pass over, except to declare that he had devoted all to the service of our lord Jesus Christ in warring with the Barbarians and in fighting for the faith. He never wished to take a wife, observing great chastity in his youth. He performed many very noble deeds in battle with the Moors, and, both for his own person and for his ability, is worthy of remembrance. The said King Don Zuane his father being near to death summoned his son the lord infante Don Heurich and commended to him the whole body of Portuguese knights praying and exhorting him to follow his holy true and laudable intent to persecute with all his power the enemies of the holy faith of Christ. And this lord in a few words pledged himself so to do. After the death of his father with the favour of King Don Dourth[3], his elder brother, who succeeded to this realm of Portogallo he waged much war in Africha against those of the Kingdom of Fess which continued for many years, the said lord striving by every possible means to destroy this Kingdom of Fess in many places. This Kingdom extends to the western sea in the region beyond the strait of Zibelterra. Thither year by year the said lord infante sent his caravels which wrought such loss to the Moors that he urged them each year to advance further. At last they reached a promontory called Capo non. This Cape has been called thus to this day and it was always the termination. Because it was found that anyone who rounded it never returned, it was called 'capo de non'—who passes never returns[4]. The said caravels at last reached this cape and did not dare to go further. So the said lord desiring to know further determined that the following year, with the favour and aid of God, his caravels should pass this Capo de non. The caravels of Portogallo being the best ships that sailed the seas and being well furnished with every necessity, he considered it possible for them successfully to sail everywhere, and being desirous to learn new things,

---

[1] Prince Henry, 'the navigator', 1394–1460.
[2] King João I, 1356–1433.   [3] King Duarte, 1391–1438.
[4] 'Chil passa ritorna nō'; the modern C. Nun.

particularly of the peoples dwelling in those lands, and to cause injury to the Moors, he had three of his caravels made ready with all necessaries, arms and munitions, as well as victuals and other things, and put courageous men on board. These set forth and passed the said cape, sailing along the coast by day and casting anchor at night. Thus they went along this coast about 100 miles beyond the Cauo de non and finding neither dwellings nor people, nothing save sandy and arid land, turned back. This lord seeing that that year he had not been able to learn anything sent his ships and men again the following year with orders to go 150 miles beyond the point these ships had reached, and further if it seemed to them possible, promising to reward them richly.

So they set forth and carried out the command of their lord but finding nothing more than sandy land and arid, without inhabitants, they returned. Notwithstanding, this lord grew every day more anxious to have knowledge of that part and persisted in despatching two the third year: and to be brief he continued to send each year until they succeeded in reaching some parts inhabited by Arabs who dwell in these deserts; and then further on a race of people called Azanagi who are brownish men of whom later there will be more mention. Thus was discovered at last the country of the first blacks and thereafter, from time to time, other races of these blacks of varied tongues, customs, and beliefs, as may be learnt more fully in the rest of this book[1].

CHAPTER II

The things discovered by D. ALOUISE DA CA DA MOSTO *in the navigation to the lands of the Blacks*

In the year of Our Lord 1454, I, Alvise da Ca' da Mosto, then aged about twenty-two years, found myself in our city of Venice. Having sailed to various parts of our Mediterranean Sea, I then determined to return to Flanders, where I had been

[1] Gil Eanes rounded Cape Bojador in 1434; Nuno Tristão reached the Gulf of Arguim in 1443.

once before, in the hope of profit: for my one thought was so to employ my youth by striving in every way possible to gain qualifications, that [with this experience of the world] I might [in later years] attain honourable distinction. When I had decided to go thither, as I have said, I made ready with what little money I had, and went aboard our fleet of Flanders galleys, which was under the command of Messer Marco Zen, knight. Thus, on the eighth of August[1] of the said year, we set forth, in God's name, from Venice. Sailing southwards, and touching at the customary ports, we ultimately made the coast of Spain.

Contrary winds delayed the galleys at Capo di San Vincenzo[2], as it is called, so that by chance I found myself at no great distance from the place where the Lord Infante Don Heurich[3] was lodged in a country estate called Reposera[4]. [Since by its remoteness from the turmoil of affairs, it was fitted for studious contemplation, he was living there very readily.] When he had news of us, he sent one of his secretaries, Antonio Conzales[5], to our galleys, accompanied by a Patrizio di Conti[6], who claimed to be a Venetian and consul of our nation in the Kingdom of Portugal—as he proved by a letter and seal from our Seignory. He was, moreover, also in the employ of the said Lord Infante. At his command they came to our galleys with samples of sugar from the Isola de Medera[7], dragon's blood[8] and other products

[1] The three galleys sailed on August 11th, 12th and 13th respectively. See p. xxxii.
[2] Cape St Vincent.
[3] Prince Henry was at this time sixty years old.
[4] Reposera was five miles inland from Sagres, on the headland of Cape St Vincent, in Algarve of which Prince Henry was Governor, and close to the old naval arsenal of Lagos. Here he had settled after his return from the capture of Ceuta in 1415, and from here he directed the work of exploration.
[5] This was presumably the Antam Gonçalves of whom we hear much from Azurara. He was one of Henry's most able captains and, in 1441, had been the first to bring home captive Azaneguys from Africa. After making several voyages he was given the governorship of Lançarote in 1447 (Azurara, II, p. 286).
[6] This Patrizio di Conti is said by Canale to have been celebrated for the extent of his geographical knowledge.
[7] Madeira.
[8] Dragon's blood is a resin obtained from the *Dracaena draco*, formerly used in medicine and now for colouring purposes.

of his domains and islands. These he displayed in my presence to many on the galleys and asked us various questions. He said that his lord had peopled newly discovered islands hitherto uninhabited, in proof of which he cited the sugar, dragon's blood, and other good and useful wares, but that this was nothing in regard to the other and greater achievements of his lord, who had for some time past caused seas to be navigated which had never before been sailed, and had discovered the lands of many strange races, where marvels abounded. Those who had been in these parts had wrought great gain among these new peoples, turning one *soldo* into six or ten.

They related so much in this strain that I, with the others, marvelled greatly. They thus aroused in me a growing desire to go thither. I asked if the said lord permitted any who wished to sail, and was told that he did, under one of two conditions. He might fit out the caravel at his own expense and load her with merchandise; on his return he would be obliged to pay to the said lord by law and custom a quarter of all he brought back keeping the remainder for himself: or the lord would at his own expense equip a caravel for whomsoever wished to go if he provided the cargo: then on the return, all that had been brought back from these parts would be halved. If nothing were brought back, then the charges would be at his expense. They said that it was impossible for anyone to return without great gain; if any of our nation wished to go thither, the said lord would receive them gladly and show them much favour, for he believed that in these parts they would find spices and other valuable products, and knew that the Venetians were more skilled in these affairs than any other nation.

Hearing this, I determined to go with them to this lord—which I did. In short, he confirmed all that they had told me, and much more, promising to do me honour and to aid me if I should desire to go. I, in truth, having learnt all this, definitely made up my mind to go, for I was young, well fitted to sustain all hardships, desirous of seeing the world and things never before seen by our nation, and I hoped also to draw from it honour and profit. Having acquainted myself with the merchandize and equipment which I should need, I returned to the galley, where I consigned to a relative[1] all I was bringing to the West, and purchased what I considered requisite for the voyage. I then disembarked, and the galleys resumed their voyage [to Flanders].

CHAPTER III

The date of the DEPARTURE *of the* GALLEYS, *and the winds with which they sailed*

My remaining at Capo di San Vincenzo greatly pleased the Lord Infante, who entertained me suitably. After many days he had a new caravel fitted out for me, of some 90 *botti* burden[2], the patron of which was one Vinzente Dies[3], of Lagus, a port 16 miles from the Capo di San Vincenzo. Furnished with all necessities, we set sail in God's name and with high hopes from San Vincenzo on the 22 March 1455. With a north-north-easterly wind abaft, we straightway set our course south-west by west for the Isola di Madeira. On the 25th of the same month, at midday we made the Isola di Porto Sancto, distant about 600 miles from Capo di San Vincenzo.

[1] His brother, Antonio. See Introduction, p. xxxi.
[2] A *botte* was a cask of wine, of varying capacity throughout the Mediterranean countries. Mr G. S. Laird Clowes considers that in this case it probably equalled half a modern ton. The figure given here refers to the carrying capacity of the caravel, and the total displacement of the vessel would be about 70 tons.
[3] Diaz is described by Azurara (II, p. 174) as a trader. He is mentioned in Azurara's account of the great expedition of fourteen sail which set out from Lagos in 1445 under the leadership of Lançarote to punish the Moors of Tider. He commanded one of the caravels and was wounded in a hand-to-hand fight with a 'Guinea' (*ibid.* pp. 178, 195).

CHAPTER IV

The island of PORTO SANCTO, *with a description of its products, and especially of 'dragon's blood' and the way it is made*

This Isola di Porto Sancto is very small, about twenty-five miles in circuit. It was discovered about twenty-seven[1] years previously by the caravels of the Lord Infante, who had peopled it with Portuguese, as it was formerly deserted. It was governed by one Bartholomio Pollastrello[2], a liege of the said lord.

The island produces corn and barley for its wants. It abounds in cattle, wild pigs, and innumerable coneys[3]. Dragon's blood is also obtained there from certain trees, that is, it is a gum which these trees yield at a certain season of the year. It is collected in this manner: the tree is gashed at the foot: the following year, at a certain time, these gashes exude the gum, which is boiled, clarified, and made into 'blood'. This tree also bears a fruit which ripens in the month of March—very good to eat, resembling a cherry, save that it is yellow. Around this island there are large fishing grounds—dory[4], and other good fish. There is no harbour, but a good anchorage, sheltered from all winds save the E.S.E. and S.S.E.: from these winds, there is little security, but, despite this, the anchorage is good. The

[1] An error for thirty-seven. The Marciana Manuscript (*Ital.* VI. 454) has seventeen years. The identity of the discoverer of Madeira and Porto Santo and the date are unknown. They appear, however, on Catalan and Italian maps of the fourteenth century. They were rediscovered in 1419–20 by João Gonçalves Zarco and Tristão Vaz Teixeira, two young squires sent out by Prince Henry to search for the land of Guinea.

[2] Bartholomeu Perestrello, a gentleman of the household of the Infant Don João, Henry's brother, whose daughter Felipa Moniz de Perestrello married Columbus. (Azurara, II, p. x.) He became Captain Donatory of Porto Santo where he imported rabbits which destroyed all the colonists' crops (*ibid.* p. 245). He made some profit from breeding goats and exporting dragon's blood (*ibid.* p. c). He accompanied Teixeira and Zarco on the voyage which led to the rediscovery of Madeira.

[3] These rabbits were all descended from a doe imported by Perestrello which gave birth to a litter before she reached Porto Santo. So prolific were her progeny that they compelled the Portuguese to abandon the island.

[4] *Sparus dentex*, according to Caddeo.

island is called Porto Sancto [because the Portuguese discovered it on All Saints' Day]. It produces I believe the best honey there is in the world, and wax, [but not] in great quantities.

CHAPTER V

MONCRICHO: *a port of the island of* MADERA

Leaving this island of Porto Sancto on 27th March, we arrived on the same day at Moncricho[1], one of the harbours of the Isola di Madera, forty miles distant. In clear weather, one is visible from the other.

CHAPTER VI

The meaning of the name MADERA, *the manner of its first settling, and its fertility*

On this Isola di Madera, the said lord twenty-four years[2] before had settled Portuguese—it had previously never been inhabited—and had appointed two of his knights governors, one of whom, Tristan Tessera[3], held the half of the island around Moncricho. The other, Zuanconzales [Zarco][4], held the half around Fonzal[5]. It is called the 'Isola de Madera', that is to say, the 'island of timber', for, when first discovered by the men

[1] Machico, the principal settlement in the northern half of the island of Madeira, the captaincy of which was given to João Gonçalves Zarco. (For the legend of the origin of its name see Azurara, II, p. lxxxv.) The southern half of the island, with Funchal as its capital, went to Tristão Vaz Teixeira (*ibid.* p. xcix), not the northern half, as stated in the text.

[2] This should be thirty years.

[3] Tristão Vaz Teixeira, who, Azurara (II, p. 213) tells us, 'had a right goodwill to serve the Infant and much desired to profit himself (for he was abundantly covetous)'.

[4] João Gonçalves Zarco, a distinguished captain who had seen much fighting and in 1445 had reached the Cape of Masts, then the farthest south. Azurara tells us, 'he was noble in all his actions' (*ibid.* II, pp. 225, 229).

[5] Funchal.

of the said Lord, there was not a foot of ground that was not entirely covered with great trees. It was therefore first of all necessary, when it was desired to people it, to set fire to them, and for a long while this fire swept fiercely over the island. So great was the first conflagration, that [I was told] this Zuanconzales, who was then on the island, was forced, with all the men, women, and children, to flee its fury and to take refuge in the sea, where they remained, up to their necks in water, and without food or drink, for two days and two nights or thereabouts, to escape destruction. By this means they razed a great part of this forest, and cleared the ground for cultivation.

There are four settlements in the island: the first Moncricho[1]; the second Santa Croce[2]; the third Fonzal[3]; the fourth Camera de Loui[4]. Although there are others, these are the principal. There must be about eight hundred men, of whom one hundred are horsemen. The circuit of the island is 140 miles: there are no landlocked harbours, but several good anchorages. It is a most fruitful and well-stocked land.

Although it is as mountainous as Sicily, yet it is very fertile: every year they harvest 30,000 Venetian *stara*[5] of corn, sometimes more, sometimes less. The soil at first yielded a return of sixty and seventy for one, but at the moment this has declined to thirty or forty for one, because the land is daily being exhausted. It is a well-watered country with copious springs and some eight streams of considerable size which flow through the island and upon which are sawmills continually working timber and planks of all kinds for the supply of all Portugal and elsewhere. Of this timber two kinds are esteemed: one of cedar, which is very odorous, and resembles cyprus. Very beautiful tables, wide and long, chests and other furnishings are made from this. The other kind is yew, which is equally attractive, rose red in colour. Since the island was so well watered, the Infante had many sugar canes planted, to his great profit. They have produced sugar to the amount of 400 *cantara*[6] at one refining, and from what I understand, they will in time produce

[1] Machico. [2] Santa Cruz. [3] Funchal.
[4] Camara de Lobos. [5] Equal to 70,000 bushels (Major, p. 249).
[6] According to Major, a *cantaro* was the same as an *alquiera*, which equalled about three gallons.

more, for the island by its warm and temperate air is most suitable for its cultivation, for it is never as cold as it can be in Cyprus or Sicily. Many pure confections of the highest standard are made from this sugar.

Wax and honey are also produced, but not in quantities: also very good wines considering the date of the settlement, sufficient for the needs of the islanders and for export overseas. Among the vines the Infante had planted are those of Maluasie[1], brought from Candia at his orders. These flourish very well in the rich, good soil, bearing almost more fruit than leaves. The bunches are very large, about four *palmi* [six, nine, and, I dare to say, even twelve inches] in length—the finest sight in the world. There are also black trellis grapes [in perfection]. They also make here very fine and excellent bows of yew which are exported to the west[2], and very fine cross-bow shafts. There are also wild peacocks, some of which are white, and partridges, but no other game save quails and numbers of wild pigs on the mountains[3]. I was told by islanders worthy of credence that at first there were vast quantities of pigeons, and that they are still to be found. They are taken in a snare, set with a stick, which traps the pigeon by the neck and knocks it from the branch. The pigeons do not know what the fowler is doing, nor do they display fear. I can well believe this because in another newly discovered island I have heard of the like. Beef is abundant, and there are many rich men, judged by the standards of the island, for it is one large garden, and everyone reaps golden rewards. There are monasteries of the order of Friars Minor on the island, men of holy lives. I have been credibly informed that [so mild is the climate both green and] ripe grapes have been seen in Holy Week [or at least in the Paschal octave].

[1] The wine, called in English Malmsey, was originally the product of the neighbourhood of Napoli di Malvasia (Monemvasia) in the Morea.
[2] I.e. western Europe.   [3] See Appendix.

CHAPTER VII

Of the CANARY ISLANDS, *ten in number, and their names*

Leaving this island of Madera, we continued our voyage southwards to the Isole de Canaria[1], a distance of about 320 miles. These islands are ten in number: seven inhabited and three deserted. These are inhabited: Lanzaroto, Forteventura, Granchamaria[2], Teneriffe, Giemera, La Palma, and Ferro. Of these seven islands four are peopled by Christians—Lanzaroto, Forteventura, Giemera, or Gomera, and Ferro. The people of the other three are idolaters. The lord of the Christian islands is Ferrara[3], knight and gentleman, native of the city of Sebillia[4], and a subject of the King of Spain. The food of these Christians, which they obtain on the islands, is barley bread, and a sufficiency of meat and milk, chiefly from the goats, of which they have many. There is neither wine nor corn save what is brought thither from other places: little fruit, and almost nothing else of value. There are plenty of wild asses, especially in the island of Ferro. The islands are forty to fifty miles apart, the one from the other, strung out in a line; and from the first to the last is roughly east to west.

[1] The earliest account of the Canaries is that given by the clerics Boutier and le Verrier in 'Le Canarien', the earliest MS. of which is *circa* 1420. Azurara's account is also slightly earlier than Cadamosto's, but the latter is more comprehensive (see Major, R. H., *The Canarian*, Hakluyt Soc., series I, vol. XLVI, and Beazley, C. R., *Geogr. Journal*, XXV (1905), p. 77).
[2] Gran Canaria.
[3] Diogo de Herrera, whose claim to the islands was based upon his marriage with Ignes Paraza, the inheritor of one of the several titles sold by Maciot de Bethencourt. They settled on Lanzarote in 1444.
[4] Seville.

CHAPTER VIII

The products of the CANARIES

Great quantities of a herb called *oricello*[1], which is used to colour bread, are exported from these islands to Cades in the river of Sibillia, whence it is shipped both to east and to west. They produce also great quantities of goatskins, large and of good quality, some tallow and good cheese.

The people of the four islands subject to the Christians are Canarians. They speak various languages and can scarcely understand each other. There are no walled towns on the islands, only villages. They have not been subdued in the mountains, for these are very high with many strong places, which the whole world could not reduce, except by siege. It is the same in all the islands held by the Christians. Each of them is large—the smallest not less than ninety miles in circuit. The other three, inhabited by heathens, are larger, and much more densely populated, particularly two, Gran Canaria, with about eight [to nine] thousand souls, and Teneriffe, the largest of the three, with about fifteen thousand souls. Palma has very few inhabitants, and is very fair to see. These three islands, being populous, have many men to defend them, are very mountainous with dangerous, fortified places, and have never been reduced by the Christians. Of Teneriffe, which is most populous, it must further be mentioned that it is one of the highest islands in the world, and in clear weather can be seen from [a great distance. A sailor worthy of credence has told me that he saw it from a distance which he estimated to be] sixty to seventy Spanish leagues, which equal 250 miles, for there is a point, [or rather a mountain] in the centre of the island like a diamond which is very high and burns continuously[2]. Christians who have been prisoners on the island affirm that from the foot to the summit of this peak is fifteen Portuguese leagues, that is, sixty of our Italian miles.

In this island there are nine lords among them, called dukes:

[1] Orchella weed, a lichen from which a purple dye is obtained.
[2] The 'Pic de Teyde' is 12,162 feet above sea level.

they are not rulers by natural law, where the son succeeds the father, but by right of the strongest. And they continually wage war among themselves, slaying each other like beasts. They have no other weapons than stones and sticks like spears, which are tipped with a sharpened horn in place of iron. Those which have no horns are sharpened at the tips, and the wood is made as hard as iron: and with these they attack. They always go naked, save some who wear goatskins, before and behind. They anoint their bodies with the fat of goats mixed with juice from some of their herbs; this thickens the skin, and keeps out the cold, although, from their southerly position, little cold prevails in these parts. They have neither walled houses nor huts of straw, but live in caves or caverns in the mountains, sustaining themselves on barley, flesh and goats' milk of which they have plenty, and on fruits, particularly figs. [As the climate is very warm] they harvest their grain in March and April.

They have no faith, nor do they believe in God: some worship the sun, others the moon and planets, and have strange idolatrous fantasies.

Their women are not held in common: but it is lawful for each to take as many as he wishes. They do not marry virgins, unless they have first lain one night with the lord, which is held to be a great honour. And if I should be asked how these things are known, my reply is that the inhabitants of the four Christian islands are wont to go by night with some of their galleys to assail these islands, and to seize these heathen Canarians, both men and women, whom they send to Spain to be sold as slaves. And it happens that at times some from these galleys are taken prisoners: the Canarians do not put them to death, but make them kill and skin goats, and prepare the meat, which they hold to be a most vile and despicable occupation[1], and they make them serve thus until they are ransomed by some means or other.

These Canarians have another custom: when their lords enter newly into office, some freely offer themselves to die in honour of the ceremony. They all go to a certain deep valley, where, after performing certain ceremonies and saying certain words, he

---

[1] This is also recorded by Azurara (II, p. 241).

who wishes to die in honour of the lord throws himself into the great valley and is dashed to pieces. The lord is then under an obligation to do the very greatest honour and benefit to all the relatives of the dead man. This rough and brutal custom is said to be thus observed, as the Christians rescued from prison affirm.

These Canarians also are lightly built, great runners and jumpers, for they are accustomed to the crags of this most mountainous island. They leap from rock to rock, barefooted, like goats, and clear jumps of incredible width. They throw stones accurately and powerfully, so that they can hit whatever they wish. They have such strong arms that with a few blows they can shatter a shield in pieces. Here I may note that I saw a Christian Canarian in the island of Madeira who wagered that, if he and three other men took twelve oranges each, he would hit each of them with his twelve oranges without their being able to touch any part of him except his hands as he caught the oranges to return them, provided they came no nearer than eight or ten paces: but no one would accept the wager because they knew that he could do even better than he said. From which I conclude that this is the most dextrous and nimble race in the world. Both men and women paint their skins green, red, and yellow with pastes of herbs, and they consider that such colours are a beautiful device, esteeming them as we do fine clothes.

And I, Alvise, was in two of these Isole di Canaria, Gomera and Ferro, which are Christian, and also at the island of Palma, but I did not land on this, as I wished to continue our voyage.

CHAPTER IX

The description of CAPO BIANCO *and the islands nearest to it*

We set sail from this island making due south towards Ethiopia; and in a few days reached Capo Blanco about 770 miles from the Canaries[1]. It is to be noted that, leaving these islands to sail

[1] Major says the distance is in fact 570 miles (p. 252).

towards this cape, one goes along the coast of Africa which is constantly on the left hand; you sail well offshore, however, and do not sight land, because the Isole di Canaria are very far out to sea to the west, each one further than its neighbour. Thus you keep a course far out from land, until you have covered at least two-thirds of the passage from the islands to Capo Bianco and then draw near on the left hand to the coast until land is sighted, in order not to run past the said Cape without recognising it, because beyond it no land is seen for a considerable distance. The coast runs back at this cape, forming a gulf which is called the 'Forna dargin'[1]. This name Dargin is derived from an islet in the gulf called Argin by the people of the country. This gulf runs in more than fifty miles, and there are three more islands, to which the Portuguese have given these names: Isola Bianca, from its sands: Isola da le Garze[2], because the first Portuguese found on it so many eggs of these sea birds that they loaded two boats from the caravels with them: the third Isola de Cuori. All are small, sandy, and uninhabited. On this Dargin there is a supply of fresh water, but not on the others.

Note that when you set out beyond the Strecto de Zibelterra [keeping this coast on the left hand, that is, of Barbary] towards Ethiopia, you do not find it inhabited by these Barbari except as far as the Cauo de Chantin[3]. From this cape along the coast to Capo Blanco commences the sandy country which is the desert that ranges on its northern confines with the mountains, which cut off our Barbary from Tunis, and from all these places of the coast. This desert the Berbers call Sarra[4]: on the south it marches with the Blacks of lower Ethiopia: it is a very great desert, which takes well-mounted men fifty to sixty days to cross—in

---

[1] Arguim, discovered in 1443 by Nuno Tristão, where a fort was erected by Prince Henry in 1448 for the protection of merchants. Its good water and safe anchorage quickly made it a valuable *entrepôt*, and it became an important trading centre. The Arab name was 'Ghir', and Azurara calls it 'Gete'.

[2] Island of Herons (Azurara, I, p. 63, and II, pp. 320–1), one of the Arguim Islands. The big expedition of 1444 rested here and refreshed themselves on the multitude of young birds.

[3] Cape Cantin, 32° 36′ N., 9° 14′ W.

[4] Sahara. The mountains are the Atlas range.

some places more, and some less. The boundary of this desert is on the Ocean Sea at the coast, which is everywhere sandy, white, arid, and all equally low-lying: it does not appear to be higher in one place than another, as far as the said Capo Bianco, which is so called because the Portuguese who discovered it saw it to be sandy and white, without signs of grass or trees whatsoever. It is a very fine cape, like a triangle, that is, on its face; it has three points, distant the one from the other about a mile.

On all this coast there are very large fisheries[1] of various and most excellent large fish without number, like those of our Venetian fisheries, and other kinds. Throughout this Forna Dargin there is little water, and there are many shoals, some of sand, others of rock. There are strong currents in the sea, on account of which one navigates only by day, with the lead in hand, and according to the state of the tide. Two ships have already been wrecked upon these banks. The aforesaid Cauo de Chantin stands approximately north-east of Capo Blanco.

You should also know that behind this Cauo Bianco on the land, is a place called Hoden[2], which is about six days inland by camel. This place is not walled, but is frequented by Arabs, and is a market where the caravans arrive from Tanbutu[3], and from other places in the land of the Blacks, on their way to our nearer Barbary. The food of the peoples of this place is dates, and barley, of which there is sufficient, for they grow in some of these places, but not abundantly. They drink the milk of camels and other animals, for they have no wine. They also have cows and goats, but not many, for the land is dry. Their oxen and cows, compared with ours, are small.

They are Muhammadans, and very hostile to Christians. They never remain settled, but are always wandering over these deserts. These are the men who go to the land of the Blacks, and also to our nearer Barbary. They are very numerous, and have

[1] The fishing fields were already being exploited under Prince Henry's licence.
[2] Wadan, an important desert market about 350 miles east of Arguim. Later, in 1487, when the Portuguese were endeavouring to penetrate the interior they attempted to establish a trading factory at Wadan which acted as a feeder to Arguim, tapping the north-bound caravan traffic and diverting some of it to the west coast.
[3] Timbuktu, see Introduction, p. xv.

many camels on which they carry brass and silver from Barbary and other things to Tanbutu and to the land of the Blacks. Thence they carry away gold and pepper[1], which they bring hither. They are brown complexioned, and wear white cloaks edged with a red stripe: their women also dress thus, without shifts. On their heads the men wear turbans in the Moorish fashion, and they always go barefooted. In these sandy districts there are many lions, leopards, and ostriches, the eggs of which I have often eaten and found good.

You should know that the said Lord Infante of Portugal has leased this island of Argin to Christians [for ten years], so that no one can enter the bay to trade with the Arabs save those who hold the licence. These have dwellings on the island and factories where they buy and sell with the said Arabs who come to the coast to trade for merchandize of various kinds, such as woollen cloths, cotton, silver, and 'alchezeli'[2], that is, cloaks, carpets, and similar articles and above all, corn, for they are always short of food. They give in exchange slaves whom the Arabs bring from the land of the Blacks[3], and gold *tiber*[4]. The Lord Infante therefore caused a castle[5] to be built on the island to protect this trade for ever. For this reason, Portuguese caravels are coming and going all the year to this island.

These Arabs also have many Berber horses[6], which they trade, and take to the Land of the Blacks, exchanging them with the rulers for slaves. Ten or fifteen slaves are given for one of these horses, according to their quality. The Arabs likewise take articles of Moorish silk, made in Granata and in Tunis of Barbary, silver, and other goods, obtaining in exchange any number of these slaves, and some gold. These slaves are brought

[1] Malaguetta pepper, see Introduction, p. xii.
[2] Probably the coarse cloth called by El Bekri in the eleventh century 'chigguiza', which was doubtless the 'shigge' purchased by Barth in Timbuktu in the nineteenth century (Barth, *Travels*, iv, p. 443).
[3] The Portuguese had now established in West Africa the insidious practice of inciting the coast tribes to raid their neighbours for slaves.
[4] The Arabic *thibr* or *tibar*, meaning gold dust.
[5] Built by Prince Henry in 1448.
[6] Leo Africanus, writing in the sixteenth century, makes several references to the trade in Barbary horses for which there was an excellent market in the Sudan. Later the Portuguese regularly shipped out horses to barter for slaves.

to the market and town of Hoden; there they are divided: some go to the mountains of Barcha[1], and thence to Sicily, [others to the said town of Tunis and to all the coasts of Barbary], and others again are taken to this place, Argin, and sold to the Portuguese leaseholders. As a result every year the Portuguese carry away from Argin a thousand slaves[2]. Note that before this traffic was organized, the Portuguese caravels, sometimes four, sometimes more, were wont to come armed to the Golfo d'Argin, and descending on the land by night, would assail the fisher villages, and so ravage the land. Thus they took of these Arabs both men and women, and carried them to Portugal for sale: behaving in a like manner along all the rest of the coast, which stretches from Cauo Bianco to the Rio di Senega and even beyond. This is a great river, dividing a race which is called Azanaghi[3] from the first Kingdom of the Blacks. These Azanaghi are brownish, rather dark brown than light, and live in places along this coast beyond Cauo Bianco, and many of them are spread over this desert inland. They are neighbours of the above mentioned Arabs of Hoden.

They live on dates, barley, and camel's milk: but as they are very near the first land of the Blacks, they trade with them, obtaining from this land of the Blacks millet and certain vegetables, such as beans, upon which they support themselves. They are men who require little food and can withstand hunger, so that they sustain themselves throughout the day upon a mess of barley porridge. They are obliged to do this because of the want of victuals they experience. These, as I have said, are taken by the Portuguese as before mentioned and are the best slaves of all the Blacks. But, however, for some time all have been at peace and engaged in trade. The said Lord Infante will not permit further hurt to be done to any, because he hopes that, mixing with Christians, they may without difficulty be converted to our

[1] Barca in Cyrenaica.
[2] According to Azurara (II, p. 288), up to the year 1448 the total number of Africans who had been carried captive to Portugal during Prince Henry's time was only 927. This passage indicates how rapidly the slave trade was increasing.
[3] The Azanaghi or Azaneguys, as Azurara calls them, were the Sanhaja, historically the most important of the Tuareg tribes, and widely distributed over the western Sahara.

faith, not yet being firmly attached to the tenets of Muhammad, save from what they know by hearsay.

These same Azanaghi have a strange custom: they always wear a handkerchief on the head with a flap[1] which they bring across the face, covering the mouth and part of the nose. For they say that the mouth is a brutish thing, that is always uttering wind and bad odours so that it should be kept covered, and not displayed, likening it almost to the posterior, and that these two portions should be kept covered. It is true that they never uncover it, except when they eat, and not otherwise for I have seen many of them.

There are no lords among them, save those who are richer: these are honoured and obeyed to some degree by the others. They are a very poor people, liars, the biggest thieves in the world, and exceedingly treacherous. They are men of average height, and spare. They wear their hair in locks down to their shoulders, almost in the German fashion—but their hair is black, and anointed every day with fish oil, so that it smells strongly, the which they consider a great refinement.

CHAPTER XI[2]

The exchange of salt for gold: and the distance it travels

That woman who has the largest breasts is considered more beautiful than the others: with the result that each woman, to increase their size, at the age of seventeen or eighteen when the breasts are already formed, places across her chest a cord, which she binds around the breasts, and draws tight with much

---

[1] The *litham*, still worn by the Tuareg; hence their name Muleththemin, meaning the Veiled People. In Roman and Byzantine times they appear not to have worn the veil, and when or why they took to it remains a problem to which no acceptable solution has been found. Its use appears always to have been restricted to the men.

[2] This and subsequent chapters are misnumbered in the original version.

force; in this way the breasts are distended, and frequent pulling every day causes them to grow and lengthen so much that many reach the navel. Those that have the biggest prize them as a rare thing.

You should know that these people have no knowledge of any Christians except the Portuguese, against whom they have waged war for [thirteen or] fourteen years, many of them having been taken prisoners, as I have already said, and sold into slavery[1]. It is asserted that when for the first time they saw sails, that is, ships, on the sea (which neither they nor their forefathers had ever seen before), they believed that they were great sea-birds with white wings, which were flying, and had come from some strange place: when the sails were lowered for the landing, some of them, watching from far off, thought that the ships were fishes. Others again said that they were phantoms that went by night, at which they were greatly terrified. The reason for this belief was because these caravels within a short space of time appeared at many places, where attacks were delivered, especially at night, by their crews. Thus one such assault might be separated from the next by a hundred or more miles, according to the plans of the sailors, or as the winds, blowing hither and thither, served them. Perceiving this, they said amongst themselves, 'If these be human creatures, how can they travel so great a distance in one night, a distance which we could not go in three days?' Thus, as they did not understand the art of

---

[1] It should perhaps be pointed out that hostilities were not provoked by the Azaneguys, or Sanhaja, but by the Portuguese who deliberately raided them to capture slaves. The first had been taken by Antam Gonçalves in 1441, only fourteen years before Cadamosto's first voyage. It was quickly realised that there was money in the enslavement of Africans and thenceforward the character of these voyages altered. Discovery was no longer pursued for its own ends, but as a source of personal gain. Buccaneers regularly descended on the north-west coast of Africa to raid the Azaneguys. All who resisted capture were ruthlessly slain. In Portugal the prisoners were sold at great profit. For the most part they were well treated and adopted the religion of their masters.

Prince Henry has been bitterly censured for permitting this disgraceful trade, which remained the curse of Africa for centuries. He believed it justified, however, as a means of saving the souls of the heathen and he should not be blamed for an attitude which accorded with the thought of his day. Moreover we have no reason to believe that he was in the least influenced by the thought of personal gain.

THE EMPIRE OF MELLI

navigation, they all thought that the ships were phantoms. This I know is testified to by many Portuguese who at that time were trading in caravels on this coast, and also by those who were captured on these raids. And from this it may be judged how strange many of our ways appeared to them, if such an opinion could prevail.

Beyond the said mart of Edon [Oden], six days journey further inland, there is a place called Tagaza[1], that is to say in our tongue 'cargador'[2], where a very great quantity of rock-salt is mined. Every year large caravans of camels belonging to the above mentioned Arabs and Azanaghi, leaving in many parties, carry it to Tanbutu[3]; thence they go to Melli[4], the empire of the Blacks, where, so rapidly is it sold, within eight days of its arrival all is disposed of at a price of two to three hundred *mitigalli*[5] a load, according to the quantity: [a *mitigallo* is worth about a ducat:] then with the gold they return to their homes.

In this empire of Melli it is very hot, and the pasturage is very unsuitable for fourfooted animals: so that of the majority which come with the caravans no more than twenty-five out of a hundred return. There are no quadrupeds in this country, because they all die, and many also of the Arabs and Azanaghi sicken in this place and die, on account of the great heat. It is said that on horseback it is about forty days from Tagaza to Tanbutu, and thirty from Tanbutu to Melli.

I enquired of them what the merchants of Melli did with this salt, and was told that a small quantity is consumed in their country. Since it is below the meridional and on the equinoctial, where the day is constantly about as long as the night, it is extremely hot at certain seasons of the year: this causes the blood to putrefy, so that were it not for this salt, they would die. The remedy they employ is as follows: they take a small piece of the salt, mix it in a jar with a little water, and drink it every

---

[1] For description of Taghaza see Introduction, p. xiv.
[2] 'A load, or charge'; other texts have 'bisaccia d'oro', i.e. wallet of gold, the gold not being obtained locally, but in exchange for salt.
[3] Timbuktu, see Introduction, p. xv.
[4] Mali, see Introduction, p. xv.
[5] One *mithqal* or *mitkal* equalled about ⅛ oz. of gold.

day. They say that this saves them. The remainder of this salt they carry away on a long journey in pieces as large as a man can, with a certain knack, bear on his head.

You must know that when this salt is carried to Melli by camel it goes in large pieces [as it is dug out from the mines], of a size most easily carried on camels, two pieces on each animal. Then at Melli, these blacks break it in smaller pieces, in order to carry it on their heads, so that each man carries one piece, and thus they form a great army of men on foot, who transport it a great distance. Those who carry it have two forked sticks, one in each hand: when they are tired, they plant them in the ground, and rest their load upon them. In this way they carry it until they reach certain waters: I could not learn from them whether it is fresh or sea water, so that I do not know if it is a river or the sea, though they consider it to be the sea. [I think however it must be a river, for if it were the sea, in such a hot country there would be no lack of salt.][1] These Blacks are obliged to carry it in this way, because they have no camels or other beasts of burden, as these cannot live in the great heat. It may be imagined how many men are required to carry it on foot, and how many are those who consume it every year. Having reached these waters with the salt, they proceed in this fashion: all those who have the salt pile it in rows, each marking his own. Having made these piles, the whole caravan retires half a day's journey. Then there come another race of blacks who do not wish to be seen or to speak. They arrive in large

---

[1] These 'certain waters' were probably the inundated area of the Niger above Timbuktu. If so, it is probable that there is a confusion between two sources of the gold, for this area is not 'a great distance' from Melli. In addition therefore to 'Wangara', in proximity to Melli, gold was probably obtained from another area farther to the south, within the region unsuited to animal transport. Professor Taylor, who has examined this question critically, concludes that this second source was the Lobi district at the headwaters of the Black Volta. The large lower lip (see below, p. 24) suggests the use of the labret. This is worn by the women of the Lobi, who collect the gold. The Lobi males take no interest in this, and gold washing has practically died out. The present inhabitants are relatively new to this district, having driven their predecessors southwards. Their tradition of gold working has therefore survived with the women absorbed by the new-comers. There is also a tradition that salt was formerly obtained from a distant source in the north. (See Taylor, E. G. R., *op. cit.*, and Labouret, H., *Les tribus du rameau Lobi*.)

boats, from which it appears that they come from islands, and disembark. Seeing the salt, they place a quantity of gold opposite each pile, and then turn back, leaving salt and gold. When they have gone, the negroes who own the salt return: if they are satisfied with the quantity of gold, they leave the salt and retire with the gold. Then the blacks of the gold return, and remove those piles which are without gold. By the other piles of salt they place more gold, if it pleases them, or else they leave the salt. In this way, by long and ancient custom, they carry on their trade without seeing or speaking to each other. Although it is difficult to believe this, I can testify that I have had this information from many merchants, Arab as well as Azanaghi, and also from persons in whom faith can be placed[1].

CHAPTER XII

How the EMPEROR *sought to take one of these traders prisoner*

Reflecting upon this, I asked the merchants how it came to be that the Emperor of Melli, who, they said, was so great a lord, had not wished so to proceed as to find out by love or by other means what people these were who did not wish to speak or to be seen. They replied that, not many years previously, an Emperor of Melli determined at all costs to get one of them in his power, and having taken counsel about it, ordered some of his men to leave a few days before the salt caravan, and proceed to the place where it was customary to pile the salt, to dig trenches near by, in which to conceal themselves. When the

[1] See Introduction, p. xvi. Valentim Ferdinand, it is interesting to note, quotes João Rodriguez to the effect that the story of the 'silent trade' was a commercial device to protect a monopoly. 'It is said that the merchants who bring the salt do not see the negroes, but leave their goods behind, for which the negroes then substitute gold. But this is not correct; for the Wangaros say that the people are unknown to them only that they alone may enjoy the profit from daring to go to them. The big-lipped negroes likewise do not wish another people to come to them; one is therefore unable to learn their requirements.' Kunstmann, F., *V. Ferdinands Beschr.* p. 270.

Blacks returned to set the gold by the salt, they were to attack them and to take two or three, whom they were to convey under close guard to Melli. To be brief, this was done. They seized four, the others taking to flight: of the four they released three, surmising that one would satisfy the desires of the lord, and not wishing to anger these Blacks more. They spoke to this man in several negro languages, but he would not reply, or speak at all, neither would he eat. He lived four days and then died. For this reason these blacks of Melli are of the opinion, after the experience they had with him who would not speak, that they are dumb. Others think that they behave thus from disdain [of doing what their ancestors had never done]. This death vexed all the blacks of Melli, for on account of it their lord could not achieve his intention. On returning to him they related the incident in due order.

Then the lord was very displeased with them, and asked what the blacks looked like. They replied they were very black in colour, with well-formed bodies, a span higher than they themselves. The lower lip, more than a span in width, hung down, huge and red, over the breast, displaying the inner part glistening like blood[1]. The upper lip was as small as their own. This form of the lips displayed the gums and teeth, the latter, they said, being bigger than their own: they had two large teeth on each side, and large black eyes. Their appearance is terrifying, and the gums exude blood, as do the lips.

Because of this incident, none of the emperor's men have since been willing to embroil themselves in similar affairs, since, as a result of the capture and death of this one negro, it was three years before the others would resume the customary exchange of gold for salt. It was thought that their lips became putrid, being in a warmer country than ours: so that these blacks, having borne much sickness and death [for this space of time], and having no other way of obtaining the salt to cure themselves, resumed the accustomed trade. On this account, it is our opinion, being unable to live without salt, they set off their plight against our action, just as the Emperor did not care

---

[1] This suggests the use of the labret, which is still worn by the Lobi women.

THE TRAFFIC IN GOLD 25

whether these blacks spoke or not, so long as he had the profit of the gold. This is what I understood from this incident, and since it is related by so many we can accept it. Because I have seen and understood such things in the world, I am one of those who are willing to believe this and other matters to be possible.

The gold thus brought to Melli is divided in three parts: one portion goes with the caravan which takes the road from Melli to a place that is called Cochia.[1] This is the route which runs towards Soria[2] [and il Cairo]: the second and third portions go with a caravan from Melli to Tanbutu. There they are separated: one portion goes to Atoet[3], whence it is carried to Tunis in Barbary through all the coast beyond: the other part goes to the above mentioned Hoden, whence it spreads towards Orā and Hona[4], towns in Barbary within the Strecto de Zibelterra, Afezes, Amarochos, Arzib, Azafi, and Amessa[5], towns in Barbary beyond the Straits. In these places it is bought by us Italians and other Christians from the Moors with the various merchandize we give them.

To return to my first subject, this is the best thing that is brought from the said land and country of the Azanaghi, that is, the brown men. Of that portion of the gold which is brought every year to Hoden, as described already, some quantity is carried to the sea coast, and sold to the Spaniards[6] who are continuously stationed on the said island of Argin for the trade of merchandize, in exchange for other things.

In this land of the brown men, no money is coined, and they have never used it. Nor, formerly, was money to be found in any of their towns. Their sole method is to barter article for article, or two articles for one, and by such means they live. It is true that I understand that inland these Azanaghi, and also the Arabs in some of their districts, are wont to employ white

[1] Gao, see Introduction, p. xvii.   [2] 'Soria' is Syria.
[3] Toet or Tuat, a group of oases in the north central Sahara on which several important caravan routes converged. At this time it was the home of a prosperous Jewish community who were the victims of a cruel massacre at the end of the century.
[4] Oran and One or Honein, the port of Tlemcen south-west of Oran.
[5] Fez, Morocco, Arzila, Saffi, and Massa, situated north of Tiznit, now no more than a ruin.
[6] 'Spain', at this period, signified the Iberian peninsula as a whole.

cowries[1], of those small kinds which are brought to Venice from the Levant. They give certain numbers of these according to the things they have to buy. I should explain that the gold they sell they give by the weight of a *mitigallo*; according to the practice in Barbary, this *mitigallo* is of the value of a ducat, more or less.

CHAPTER XIII

The respect paid to rich men; the clothes of their women; and their weapons

Those who inhabit this desert have no religion, nor any natural king. It is true that they recognize and do reverence to one more than to another, and the richer are more sought after, as in many other places; thus they have greater followings of people, but they are not lords. The women of this country are light brown, and those of higher rank are wont to wear coarse cotton cloth, which comes from the land of the Blacks, and some the above mentioned headgear, to which they give the name of Alchizel[2]. They do not wear shifts.

They ride horses in the Moorish fashion: but they have not many, since, the country being sterile, they cannot support them, and also because of the great heat, they do not live long. The districts of this desert are very hot; this heat, and the lack of water, make the land arid and infertile. In these parts it never rains except for three months in the year, August, September, and October.

I have also seen in these parts that in certain years very great numbers of locusts appear; they are like grasshoppers, but larger, and red in colour: [they are like the grasshoppers which breed and jump in the meadows, but larger, and red and yellow in colour]. They appear in the air at certain seasons in such

---

[1] According to El Bekri cowry shells were used as currency in Mauritania in the eleventh century. Probably then, and certainly during succeeding centuries, they were imported into Western Africa from Cairo.

[2] See note 2, p. 17 above.

numbers that they cover it so that the sun cannot be seen. As far as one can see, from twelve and more miles around, all is covered by these animals, both air and earth. This is a stupendous sight. Wherever they settle, nothing remains above ground, for they destroy everything. [To be visited by them is considered to be a great plague.] If they were to come every year, it would be impossible to dwell in this land—but they do not come more than once in every three or four years. On one occasion when I was passing through this land I saw them on the coast—their numbers were incalculable.

CHAPTER XIV

The RIO DE SENEGA, *which divides the desert from the fertile land*

When we had passed in sight of this Cauo Bianco, we sailed on our journey to the river called the Rio de Senega, the first river of the Land of the Blacks, which debouches on this coast. This river separates the Blacks from the brown people called Azanaghi, and also the dry and arid land, that is, the above mentioned desert, from the fertile country of the Blacks. The river is large; its mouth being over a mile wide, and quite deep. There is another mouth a little distance beyond, with an island between. Thus it enters the sea by two mouths, and before each of them about a mile out to sea are shoals and broad sand-banks. In this place the water increases and decreases every six hours, that is, with the rise and fall of the tide. The tide ascends the river more than sixty miles, according to the information I have had from Portuguese who have been [many miles] up it [in caravels]. He who wishes to enter this river must go in with the tide, on account of the shoals and banks at the mouth. From Cauo Bianco it is 380 miles to the river: all the coast is sandy to within about twenty miles of the mouth. It is called Costa de Antte rotte, and is of the Azanaghi, or brown men.

It appears to me a very marvellous thing that beyond the river all men are very black, tall and big, their bodies well formed; and the whole country green, full of trees, and fertile: while on this side, the men are brownish, small, lean, ill-nourished, and small in stature: the country sterile and arid. This river is said to be a branch of the river Nile, of the four royal rivers: it flows through all Ethiopia, watering the country as in Egypt: passing through 'lo caiero'[1], it waters all the land of Egypt. This river has many other very large branches, in addition to that of Senega, and they are great rivers on this coast of Ethiopia, of which more will be related later[2].

[1] Cairo.
[2] This passage appears somewhat expanded in Ramusio:
'This river, according to what learned men say, is a branch of the river Gihon, which flows from the terrestrial paradise. This branch was called by the ancients, Niger, which waters all Ethiopia: and drawing near to the Ocean Sea towards the west where it debouches, it forms many other branches and rivers, as well as this Senega. Another branch of the Gihon is the Nile, which flows through Egypt, and falls into our Mediterranean Sea. And this is the opinion of those who have known the world.'

This is an example of the difficulties encountered by renaissance students in attempting to fit observed facts into the framework of traditional geography. The four rivers are the Gyon (identified with the Nile), the Tigris, the Euphrates, and the Phison, which were considered by medieval cosmographers to have their common source in the terrestrial paradise, placed in the East. Some classical geographers had made the Niger a western tributary of the Nile. The confusion between the westward flowing Senegal and the Niger led to these two rivers being combined to give the 'Nile' a second outlet in the Atlantic, as set out above. It is strange that Ramusio should have accepted this account, for fifteenth-century cartographers, e.g. Fra Mauro and the draughtsman of the Estense map of 1450, distinguish a number of westward flowing rivers from the Niger, and in Diogo Gomes's time, a more correct idea of the hydrography had been gained. (See below, p. 93.) Ramusio was no doubt misled by Leo Africanus's statement that the Niger flowed westwards to the ocean.

CHAPTER XV

The LORDS *who rule on the* COAST *of* CAPO VERDE

The country of these first blacks of the Kingdom of Senega is at the beginning of the first Kingdom of Ethiopia. It is all low-lying country, and many people live on the banks of this river. They are called Zilofi[1]. For a great distance beyond, it is low country, and beyond the river likewise, except for Cauo Verde, which is the highest land on all this coast, for 400 miles beyond this Cauo Verde, and for 900 miles on this side of the said cape, the whole coast is flat. [and the people who dwell along its banks are called Gilofi. And all this coast and the known country behind is all lowland as far as the river, and also beyond this river to Capo Verde. This Cape is the highest land on the whole coast, that is for four hundred miles beyond the said Cape.]

The King of Senega in my time was called Zuchalin [Zucolin —a youth of twenty-two years. This Kingdom does not descend by inheritance:] but in this land there are divers lesser lords, who [three or four of whom,] through jealousy, at times agree among themselves, and set up a King of their own, if he is in truth of noble parentage. This King rules as long as is pleasing to the said lords [that is, according to the treatment they receive from him]. Frequently [they banish him by force: and as frequently] the King makes himself so powerful that he can defend himself against them. Thus his position is not stable and firm, as is that of the Soldan of Babilonia[2]: but he is always in dread of deposition [death or exile].

You must know that this King is lord of a very poor people, and has no city in his country, but villages with huts of straw only. [They do not know how to build houses with walls:] they

---

[1] The Jalof, or Wolof, a branch of the true negro race of Africa. 'They are said to be the blackest and most garrulous of African peoples (their name being variously explained as meaning 'speaker' or 'black').' Seligman, C. G., *Races of Africa*, p. 59. The accuracy of Cadamosto's description of the negroes is substantiated by later writers.

[2] Cairo, or perhaps Baghdad.

have no lime with which to build walls, and there is great lack of stones. This Kingdom, also, is very small; it extends no more than two hundred miles along the coast, and, from the information I had, about the same distance inland or a little more. The king lives thus: he has no fixed income [from taxes]: save that each year the lords of the country, in order to stand well with him, present him with horses, which are much esteemed owing to their scarcity[1], forage, beasts such as cows and goats, vegetables, millet, and the like. The King supports himself by raids, which result in many slaves from his own as well as neighbouring countries. He employs these slaves [in many ways, mainly] in cultivating the land allotted to him: but he also sells many to the Azanaghi [and Arab] merchants in return for horses and other goods, and also to Christians, since they have begun to trade with these blacks.

The King is permitted to have as many wives as he wishes, as also are all the chiefs and men of this country, that is, as many as they can support. Thus this King has always thirty of them, though he favours one more than another, according to those from whom they are descended. This is his manner of living with his wives: he has certain villages and places, in some of which he keeps eight or ten of them. Each has a house of her own, with young servants to attend her, and slaves to cultivate the possessions and lands assigned by the lord, [with the fruits of which they are able to support themselves]. They have also a certain number of beasts, such as cows and goats, for their use; in this way his wives have the land sown and the beasts tended, and so gain a living. When the King arrives at one of these villages, he goes to the house of one of his wives, for they are obliged to provide, out of this produce, for him and those accompanying him. Every morning, at sunrise, each prepares three or four dishes of various foods, either meat, fish, or other Moorish foods according to their practice. These are sent by their slaves to be put at the disposal of their lord, so that within an hour forty or fifty dishes are assembled; when the time at which the lord wishes to eat has arrived, he picks out whatever tempts him, and gives the remainder to those in his train. But

[1] See note 3, p. 35 below.

he never gives his people abundance to eat, so that they are always hungry. In this fashion he journeys from place to place without giving any thought to his victuals, and lodges sometimes with one wife, sometimes with another, so that he begets numerous sons, for when one is pregnant he leaves her alone. All the other chiefs of this country live in this same fashion.

CHAPTER XVI

The CUSTOMS *of the* BLACKS, *and their* BELIEFS

The faith of these first Blacks is Muhammadanism: they are not however, as are the white Moors, very resolute in this faith, especially the common people. The chiefs adhere to the tenets of the Muhammadans because they have around them priests of the Azanaghi or Arabs, [who have reached this country]. These give them some instruction in the laws of Muhammad, enlarging upon the great disgrace of being rulers and yet living without any divine law, and behaving as do their people and lowly men, who live without laws; and since they have converse with none but these Azanaghi and Arab priests, they are converted to the law of Muhammad. But since they have had converse with Christians, they believe less in it, for our customs please them, and they also realise our wealth and ingenuity in everything as compared with theirs. They say that the God, who has bestowed so many benefits, has shown his great love for us, which could only be if his law were good—but that, none the less, theirs is still the law of God, through which they will find salvation, as we through ours.

These people dress thus: almost all constantly go naked, except for a goatskin fashioned in the form of drawers, with which they hide their shame. But the chiefs and those of standing wear a cotton garment—for cotton grows in these lands. Their women spin it into cloth of a span in width. They are unable to make wider cloth because they do not understand how to card it for weaving. When they wish to make a larger piece, they sew four or five of these strips together[1]. These garments are made to reach half way down the thigh, with wide sleeves to the elbow. They also wear breeches of this cotton, which are tied across, and reach to the ankles, and are otherwise so large as to be from thirty to thirty-five, or even forty *palmi* round the top; when they are girded round the waist, they are much crumpled and form a sack in front, and the hinder part reaches to the ground, and waggles like a tail—the most comical thing to be seen in the world. They would come in these wide petticoats with these tails and ask us if we had ever seen a more beautiful dress or fashion: for they hold it for certain that they are the most beautiful garments in the world. Their women, both married and single, all go covered with girdles, below which they wear a sheet of these cotton strips bound across, half way down their legs. Men and women always go barefoot. They wear nothing on their heads: the hair of both sexes is fashioned into neat tresses arranged in various styles, though their hair by nature is no longer than a span. You must know also that the men of these lands perform many women's tasks, such as spinning, washing clothes and such things. It is always very hot there, and the further one goes inland, the greater the heat: by comparison, it is no colder in these parts in January than it is in April in our country of Italy.

CHAPTER XVII

Men clean in their persons and filthy in eating

The men and women are clean in their persons, since they wash themselves all over four or five times a day: but in eating they are filthy, and ill-mannered. In matters of which they have no experience they are credulous and awkward, but in those to which they are accustomed they are the equal of our skilled men. They are talkative, and never at a loss for something to say: in

---

[1] This is still a common practice in the western Sudan. See note 2, p. 17 above.

general they are great liars and cheats: but on the other hand, charitable, receiving strangers willingly, and providing a night's lodging and one or two meals without any charge[1].

CHAPTER XVIII

How the LORDS *of the* BLACKS *of the* KINGDOM *of* SENEGA[2] *fight*

These negro chiefs are continually at war, the one with the other, and also frequently with their neighbours. They wage their wars on foot, for they have very few horses, as these cannot live on account of the great heat, as I have already said. They do not wear armour, for they have none, save round, broad shields [which they make from the skin of an animal called *danta*[3], which is very hard to penetrate:] for the attack they carry numerous 'Azanage', which are their spears. They hurl these very swiftly, for they are great masters at throwing them. These darts have a tip of iron wrought with barbs, made in various styles: so that when they strike, it lacerates the flesh to withdraw them. They also carry some Moorish weapons, in the style of a short scimetar, that is, curved: they are made of iron, not of steel, for they obtain iron from the kingdom of Gambra of the Blacks beyond, but they cannot make steel. If there is iron in their land, they do not know of it, or are not skilled in working it. They carry also another weapon, a kind of lance similar to our javelin, but they have no other arms. Their combats are very fatal; since their bodies are unprotected, many are slain. They are very courageous and brutal, for in danger

---

[1] A striking and no less pleasing feature of the history of the exploration of western Africa is the frequent tributes paid by the great explorers, irrespective of race or creed, to the hospitality of the negroid peoples of the western Sudan. Ibn Battuta, Leo Africanus, Mungo Park and Henry Barth all join with Cadamosto in remarking upon the kindliness of these people, which was in such marked contrast to the barbarity of the light-skinned desert tribes.

[2] Misprinted 'Gambra' in the *Paesi*.

[3] Addax gazelle, called by Leo Africanus 'Elamt' (*The History and Description of Africa*, II, p. 309).

they prefer to be killed rather than to seize the opportunity of fleeing. They are not terrified at seeing their companions fall, as though, being accustomed to this, they are not grieved by it; and they have no fear whatever of death.

CHAPTER XIX

Of the great SWIMMERS *and of the* KINGDOM *of* SENEGA

They have no ships: nor had they seen any from the beginning of the world until they had knowledge of the Portuguese. It is true that those who live on the banks of this river, and others along the sea coast, have canoes [called *almadie*, made from tree-trunks], the larger of which carry three or four men. In these they constantly fish, ferry across the river, or paddle from place to place. These same Blacks are the most expert swimmers in the world, as I know from my experience of the peoples of these parts. According to what I have been able to learn, this Kingdom of Senega marches inland on the east with the country known as Tuchuror[1]; on the south with the Kingdom of Gambra: on the west with the Ocean Sea: and on the north with the already mentioned river, which divides the Tawny men from these first Blacks.

About five years before I made this voyage, this river was discovered by three caravels of the lord Infante[2]. These ascended it and made peace with the Blacks, with whom they commenced to trade. Thus ships were lying there from year to year down to my visit.

[1] Tucolor or Tekrur, a negro kingdom on the Senegal, first mentioned as far back as the eleventh century. The Tucolors were eventually driven by Berber invaders from the desert southward across the Gambia to their present home in Futa, where they are now known as Takarir.

[2] The Senegal river, which had been discovered by Dinis Dias in 1445, that is to say ten years before Cadamosto's visit, not five as he says. Perhaps for 'discovered' we should read 'explored'. The Portuguese regarded the Senegal as the frontier separating the Sanhaja (Azaneguys) or Tawny Moors from the Negroes or Black Moors.

CHAPTER XX

The TRADE between ALOUISE DA MOSTO and the LORD BUDOMEL

I passed beyond this river of Senega in my caravel and sailed to the country of Budomel,[1] fifty miles[2] by the coast from the said river: all this coast is low, without mountains. This name Budomel is the title of the ruler [not the name of the country]. It is called 'Terra de Budomel', that is to say the land of that lord, or count.

At this place I made my caravel fast, in order to have converse with this ruler, for certain Portuguese who had had dealings with him had informed me that he was a notable and an upright ruler, in whom one could trust, and who paid royally for what was brought to him. Since I had with me some Spanish horses, which were in great demand in the country of the Blacks[3], not to mention many articles such as woollen cloth, Moorish silk[4] and other goods, I made up my mind to try my fortune with this lord.

Accordingly, I cast anchor at a place on the coast of this country, called 'le Palme de Budomel', which is a roadstead, not a port. This done, I caused my negro interpreter to announce my arrival, with horses and goods for his service if he had need of them. To be brief, this lord, being informed of this, took horse and rode down to the sea-shore, accompanied by fifteen horsemen and one hundred and fifty footmen. He sent to me to say that it would please him if I would go ashore to see him, and that he would treat me with honour and esteem. Having heard of his high reputation I went thither. He entertained me

[1] The elective sovereign of Cayor is known as the Damel. Cayor formerly extended over a considerably greater area to the north-east of Cape Verde than is designated by the name at present. The inhabitants are Jalof.
[2] 80 miles, in Codex Marciana (*Ital.* VI. 485).
[3] Horses were a regular article of the trans-Saharan caravan trade. Leo Africanus tells us that in Gao, the Songhai capital, which he visited about 1513, horses bought in Europe for ten ducats and transported across the desert were selling for forty and sometimes fifty ducats (Leo, III, p. 827).
[4] Silk had long been an important article of trans-Saharan trade.

to a great feast, and after much talk I gave him my horses, and all that he wished from me, trusting to his good faith. He besought me to go inland to his house, about two hundred and fifty [twenty-five] miles from the shore. There he would reward me richly, and I might remain for some days, for he had promised me 100 slaves in return for what he had received. I gave him the horses with their harness and other goods, which together had cost me originally about three hundred ducats. I therefore decided to go with him, but before I left he gave me a handsome young negress, twelve years of age, saying that he gave her to me for the service of my chamber. I accepted her and sent her to the ship. My journey inland was indeed more to see interesting sights and obtain information, than to receive my dues[1].

CHAPTER XXI

How ALOUISE DA MOSTO *went inland with the* LORD BUDOMEL

I accompanied Budomel into the interior, being provided by him with horses, and all I required. When we had approached within perhaps four miles of his dwelling, he placed me in the charge of a nephew of his, called Bisboror, the lord of a village at which we had arrived. He took me into his house, and rendered me all honour and good company, and I remained there about 28 days in the month of November[2]. During this time I frequently visited the lord Budomel, always in the company of his nephew.

I also saw much of the manner of life in this land, to which I shall refer later. I had the more occasion to see it for it was necessary to return by land to the river of Senega, for there was so much bad weather on this coast that I was obliged when I wished to embark to have my ship brought round to this river, and to go thither myself by land. Among the things I witnessed

[1] It was this spirit which distinguished Cadamosto from his contemporaries, with whom commercial gain seems generally to have been uppermost.
[2] This appears to be an error, as the next month mentioned is June.

was an incident when I wished to send a letter to those on board, warning them that to take me off they must put into this river, whither I was going by land. I asked these blacks if there was anyone who could swim well and was bold enough to carry my letter to the ship, three miles off shore. At once many said they were. As the sea was high and there was much wind, I said it appeared to me almost impossible that any man could succeed—principally because at a bowshot from the shore there were sandbanks and likewise farther out to sea at another bowshot more banks, between which the current was so strong, now rising, now falling, that it would be very difficult for any swimmer to keep from being swept away. The sea was breaking so heavily over these banks that it appeared impossible to pass them. However that might be, two negroes presented themselves as willing to go. I asked them what I must give them, to which they replied 'Two maoulgis [maiulie] of tin each', which are worth a grosson each. Thus for this reward each agreed to take my letter to the ship, and entered the water.

I cannot narrate the difficulties they encountered in attempting to pass those banks in that sea. Whenever they disappeared from sight for a considerable time I thought that they must have been drowned. At last one could no longer withstand such buffeting from the waves, so many were breaking over him, and turned back: but the other stood firm, and fought for a long hour on the bank. In the end, he crossed it, and bore the letter to the ship, returning with a reply. This was to me a marvellous action, and I concluded that these coast negroes are indeed the finest swimmers in the world.

This is what I was able to observe of this lord and his manners, and his house. First, I saw clearly that, though these pass as lords, it must not be thought that they have castles or cities, as I have already explained. The King of this realm had nothing save villages of grass huts, and Budomel was lord only of a part of this realm—a thing of little account. Such men are not lords by virtue of treasure or money, for they possess neither, nor do they expend any money: but on account of ceremonies and the following of people they may truly be called lords: indeed they receive beyond comparison more obedience than our lords.

CHAPTER XXII

The HOUSES and VILLAGES of BUDOMEL and his many WIVES

It must be understood that the dwelling of such a King is never fixed: he has a number of villages to support his wives and families. In the village where I was, which was called 'Casa sua', there were from forty to fifty grass huts close together in a circle, surrounded by hedges and groves of great trees, leaving but one or two gaps as entrances. Each hut has a yard divided off by hedges, and thus one goes from yard to yard, and from house to house.

In this place Budomel had nine wives: and likewise in his other dwellings, according to his will and pleasure. Each of these wives has five or six young negro girls in attendance upon her, and it is as lawful for the lord to sleep with these attendants as with his wives, to whom this does not appear an injury, for it is customary. In this way the lord changes frequently.

These negroes, both men and women, are exceedingly lascivious: Budomel demanded of me importunately, having been given to understand that Christians knew how to do many things, whether by chance I could give him the means by which he could satisfy many women, for which he offered me a great reward. They are also very jealous, and allow no one to enter the huts where their wives live—not even trusting their own sons.

CHAPTER XXIII

The HOUSEHOLD always in attendance on BUDOMEL

This Budomel always has at least two hundred negroes with him, who constantly follow him, though it is true they are always coming and going: and in addition there is no lack of people who come to meet him from divers places. At the entrance of the house, before one reaches his living and sleeping quarters, there are seven large and enclosed courtyards, leading from one to another. In the midst of each is a great tree, so that those in attendance may sit in the shade. His household are divided among these courts according to their rank, so that in the first by the entrance are those of least account, and the nearer one approaches the apartment of Budomel, the greater is the dignity of those living in these courts, up to the door of Budomel. Very few men are bold enough to penetrate thither, save Christians, who are allowed to go about freely when they are present, and also the Azanaghi priests, that is those who are learned in the law—for more liberty is bestowed upon these two classes than upon his own negro subjects.

CHAPTER XXIV

The CUSTOMS of BUDOMEL and of those who HONOUR and SALUTE HIM

This Budomel exhibits haughtiness, showing himself only for an hour in the morning, and for a short while towards evening. At these times he places himself in the court before the door of the first dwelling, to which as I have said none save the two classes mentioned and some men of standing are admitted.

Such lords as he, when granting audience to anyone, display much ceremony: however considerable he who seeks audience may be, or however high born, on entering the door of Budomel's courtyard he throws himself down on his knees, bows his head to the ground, and with both hands scatters sand upon his naked shoulders and head. This is their manner of greeting their lord. No man would be bold enough to come before him to parley, unless he had stripped himself naked save for the girdle of leather they wear. The client remains in this posture for a good while, scattering sand over himself; then, without rising, but grovelling on hands and knees, he draws nearer. When within two paces, he begins to relate his business, without ceasing to scatter sand, and with head bowed as a sign of the greatest humility. The lord scarcely deigns to take notice of him, continuing to speak with others: then, when his vassal has done, he replies arrogantly in few words: thus by this act he shows much haughtiness and reserve: if God himself came to earth I do not think that they could do Him greater honour and reverence.

All this appears to me to proceed from the great fear and dread in which these people hold their lord, since for the most trivial misdeed he seizes and sells their wives and children. Thus it appears to me that his power exacts obedience and fear from the people by selling their wives and children. [In two ways they exercise the rights of lords, and display power, that is, in maintaining a train of followers, in allowing themselves to be seen rarely, and in being greatly reverenced by their subjects.]

CHAPTER XXV

The MOSQUE *of* BUDOMEL, *and the manner of* WORSHIP *and of* LIVING

Through the great familiarity which Budomel showed me, I was permitted to enter the mosque where they pray: arriving towards evening, and having called those of his Azanaghi, or Arabs, who are constantly on duty in the mosque—we would call them priests (they are those who are learned in the laws of Muhammad)—he entered with some of his chief lords into a certain place [where the mosque stood. There they prayed in this fashion:] standing upright and frequently looking up to the sky, [they took two paces forward, and recited some words in a low voice:] then bowed down very often and kissed the earth. Everything that the priest did, the lord and the others did also; standing upright and bowing to the earth [ten or twelve times]. And thus they continued for the space of half an hour.

When he had finished, he asked me what I thought of it. As he was very anxious to hear the articles of our faith recited, he frequently asked me if I would consent to repeat some of them for him. Finally I told him that his faith was false, and that those who had instructed him in such things were ignorant of the truth. On many grounds I proved his faith to be false and our faith to be true and holy—thus getting the better of his learned men in argument.

The lord laughed at this, saying that our faith appeared to him to be good: for it could be no other than God that had bestowed so many good and rich gifts and so much skill and knowledge upon us [but that he had not given us good laws]. He, on the contrary, had given them good laws, and he considered it reasonable that they would be better able to gain salvation than we Christians, for God was a just lord, who had granted us in this world many benefits of various kinds, but to the negroes, in comparison with us, almost nothing. Since he had not given them paradise here, he would give it to them hereafter.

In this he showed good powers of reasoning and deep understanding of men. He was much pleased with the actions of the Christians, and I am certain it would have been easy to have converted him to the Christian faith, if he had not feared to lose his power, for his nephew, in whose hut I lodged, often told me so. He himself was exceedingly willing that I should explain our laws to him, saying that it was good to listen to the word of God.

In his way of living [that is, of eating] he conducts himself similarly to the King of Senega, as I have described above. Each of his wives sends him a certain number of dishes of food every day. All the negro lords and men of this land follow this fashion, their women supplying them with food. They eat on the ground, like animals, without manners. No one eats with these negro rulers, save those Moors who teach the law, and one or two of their chief men. All the other lesser people eat ten or twelve together, all helping themselves from a dish of food placed in their midst. They eat little at a time, but frequently, some four or five times a day.

CHAPTER XXVI

The GRAIN *and the* WINE *which are produced in the* KINGDOM *of* SENEGA

No corn, rye, barley, spelt[1], or vines grow in this Kingdom of Senega, nor from thence onwards, in any regions of the land of the Blacks. This is because the country is very hot and without rain for nine months in the year, that is from October to the end of June. Although they have attempted to sow these grains [which they have obtained from us Christians], they will not grow because of the great heat. [Corn requires a temperate land and frequent rain, which this country lacks.] It appears that they grow various kinds of millet, small and large, beans, and kidney beans, which are the largest and finest in the world. The kidney beans are as big as the long hazel nuts familiar to us, all spotted with different colours, as though painted, and very beautiful to the eye. The beans are broad, thin, and of a bright red colour, though there are others white in colour, very beautiful[2]. They are sown in the month of July, and harvested in September, because during the period of the rains they till the land. They sow and harvest within the space of three months. They are very bad labourers—unwilling to exert themselves to sow more than will barely support them throughout the year. Few trouble to raise supplies for market.

Their method of working is as follows: four or five of them take their places in the field equipped with certain small spades fashioned like mattocks, and advance throwing the soil before them, a practice contrary to that of our labourers who when tilling the soil draw it towards them. These throw it forward with their mattocks, and do not penetrate more than four inches or so. This is their method of agriculture, and since the ground is fertile and rich, it brings forth those things described above. They drink water, milk, or palm wine. This wine is a liquid which flows from a tree similar to, but not identical with, that which bears dates. There are not many of

[1] A kind of wheat.   [2] See Appendix.

these trees: they give forth this liquid, which the negroes call 'mignol', almost the whole year. They proceed in this manner: they make three or four gashes at the foot of the tree, from which flows a brownish liquid like the whey of milk, and place gourds beneath to collect it. A tree does not yield a large quantity, about two gourds in a day and a night. It is very good to drink, and is as intoxicating as wine if water is not added. When first collected it is as sweet as the sweetest wine in the world, but day by day it looses its sweetness and becomes sour. It is better after three or four days than at first. I drank it many times during my sojourn in this country, and preferred it to our own. There is not a sufficient quantity of this *mignol* for everyone to have it in abundance, but all have a reasonable amount, the chiefs the greatest. The trees which yield this wine are common to all [that is, they are not the property of a particular person], for they do not have orchards, or own such trees individually. They are all in the forest, common to everyone to tap and to avail himself of the liquor.

CHAPTER XXVII

FRUITS *of various kinds, and a marvellous* OIL

There are fruits of various kinds, similar to ours, and also others, which are good and which they eat. All are forest trees, that is, wild, for they do not cultivate orchards as we do, and I think that if they were cared for, as ours are, they would bear good and perfect fruit [for the air and soil are excellent].

His country is all productive land, with good pastures, and innumerable great and very beautiful trees, unknown to us. There are also many lakes of sweet water, not large, but very deep, in which many good fish, of other species than ours, are to be found, also many water serpents, called cockatrice.

In this country they use a certain oil in their food, [the making of which I do not know]. It has three properties, the scent of violets, the taste of our olive oil, and a colour which tinges the food like saffron, but is more attractive[1].

There is also to be found in this country a species of tree which bears red nuts with black 'eyes' in great quantities—but they are small[2].

CHAPTER XXVIII

GREAT SERPENTS *which can* SWALLOW A GOAT, *and* CHARMERS *of them*

There are divers sorts of animals, and large numbers of snakes, great and small: some are poisonous, others not. Of the bigger, some are two paces and more in length, but without wings or feet which serpents [are said to] possess. They are so large that snakes are found which swallow a goat whole, without tearing it to pieces. It is said that these great ones are found in swarms in some parts of the country, where there are also enormous quantities of white ants, which by instinct make houses for these snakes with earth which they carry in their mouths. When these are made they resemble ovens. Of these houses they make a hundred or a hundred and fifty in one spot, like fine towns.

These negroes are very great charmers of all things, and especially of these serpents. I have heard it said by a Genoese, a man worthy of credence, that having found himself the year previous to me in the country of this Budomel, and being asleep one night in the house of his nephew, Bisboror, where I lodged, he heard in the middle of the night much hissing outside the house. On rousing himself, he saw that the said Bisboror had arisen, and having summoned two of his negroes was preparing to mount a camel and to depart. The Genoese demanded of him whither he wished to go at such an hour. He

---

[1] Probably ground-nut oil.
[2] Probably kernels from the oil palm.

replied it was on one of his duties—and suddenly made off. After an interval he returned to the house. The Genoese again questioning him, he replied: "Did you not hear hissing a while ago—hisses outside the house?" The Genoese replying in the affirmative, this is what Bisboror told him: "They were serpents, which, if I had not gone to perform a certain incantation which we of these parts employ, and by which I have turned them away, would have slain many of my animals this night."

The Genoese said that he marvelled at such a thing, for no Christian would credit it. Bisboror replied that there was nothing marvellous in it, because his uncle Budomel could work much more powerful spells—for when he wished to collect venom to poison his weapons he made a great circle into which by a charm he drew all the snakes of the surrounding country. That which appeared to him to be the most poisonous he killed with his own hands, the rest he allowed to go. Squeezing out the blood, he mixed it with the juice from a certain tree, which I have seen, and prepared a compound with which he poisoned his weapons. A wound from these, drawing a little blood, although it might be small, would kill the wounded person in a quarter of an hour. This Genoese told me that Bisboror wished him to see the incantation put to the proof, but that he did not care to learn more.

Thus I conclude that all these negroes are great magicians; [and others could bear witness to the truth of this charming of the snakes[1], because I am given to understand] in these our parts of Christendom people are also to be found who can charm snakes.

[1] Snake charmers are still commonly found in the principal market-places of western Africa.

CHAPTER XXIX

Great numbers of WILD ANIMALS, *especially* ELEPHANTS *and* GIRAFFES

In this Kingdom of Senega there are no domestic animals other than bulls, cows, and goats: no sheep are bred, nor could they live there on account of the great heat [for the sheep is an animal which prefers the temperate climes, and would sooner live in cold countries than in hot: therefore the Lord our God has in this world provided for each according to his wants: we who dwell in the cold, would be unable to live without wool: and to these negroes who are born in the hot climes, and have no need of garments, God has given, not sheep, but cotton]. The cows and bulls of this country [and also of all the land of the Blacks,] are very much smaller than ours. [I believe this is a result of the heat.] Occasionally one sees a cow with a red hide: most are black or white, or black and white mixed. Wild beasts of prey are lions and lionesses, and leopards in great numbers, also wolves[1], deer, and hares. There are also wild elephants, though they do not domesticate them, as in other parts of the world. These animals go in herds, as our swine do in the forests. [Of their size I shall say nothing, for I believe that everyone must know that the elephant is an animal with a very large body and short limbs. His size may be gathered from the teeth of ivory which are brought to our parts.] These elephants have two large teeth at each side of their mouths, [that is, one on each side,] like the wild boar, [but set in the lower jaw[2]. There is no other difference,] save that the points of the teeth of the boar are turned upwards, while those of the elephant are turned downwards towards the ground. [I must explain that these animals have knees, which they bend when walking. I say this because I have heard others who have formerly been in these parts declare that the elephant is unable to kneel down; and that they sleep standing on their feet. This is a great lie, for they lie down on the ground and rise up like any other animal.]

---

[1] ? Jackals or hyenas.   [2] This, of course, is incorrect.

WILD ELEPHANTS

They never shed their great tusks, until death. The elephant is an animal that does not attack man unless man attacks him. The method of the elephant in attacking a man is to overtake him and deal him so strong an upward blow with his long trunk of a snout, which is like a kind of very long nose, and which he can withdraw and extend as he pleases, that the man falls to the ground as though hit by a bolt from a cross-bow. There is no man so swift that the elephant cannot overtake him in open country, the elephant going at a swift walk only, since for their great size they have a very rapid pace. They are much more dangerous when with their young than at other times. They do not have more than three or four at a time[1]. They live upon leaves and fruits from the trees, which they break down on great branches with their snouts, which are very big and strong. The trunk is on the lower jaw, and they can lengthen and shorten it at will. With it they gather all their food, and drink water, putting it in their mouths, which are in their breasts. They frequent thick woods, where they wallow in the marshes like swine.

I understand that in these lands there are giraffes, and other animals of the most savage kinds.

CHAPTER XXX

PARROTS *and* BIRDS *of various kinds*

The birds in these parts are of many species, mainly parrots in great numbers, which fly everywhere. The negroes dislike them intensely, for they damage the millet and vegetables in their fields. There are [not, as is said,] many varieties of them, but I have only seen two:—large and small, and they are of divers colours—green, brownish and yellow. [One, similar to those brought from Alessandria, only apparently somewhat smaller. The other kind are much larger, with a brownish head, neck, beak, and feet, and a yellow and green body.][2] I took many from their nests, of which a large number died, but I brought

---

[1] Cow elephants usually have only one calf.   [2] See Appendix.

many others back to Spain, above one hundred and fifty, selling them for half a ducat each. These parrots are very industrious in building their nests, which they make with rushes, round like a ball, in this fashion: settling on a palm, or any other tree with branches as thin and as fragile as possible, they weave a rush from the tip of the branch which hangs down two handbreadths and at the end of this they build their nest, marvellously woven, so that when it is completed, it remains a ball hanging from the rush, in which there is only one hole for entrance[1]. They build thus so that the snakes, who eat the young birds, are unable to go on to the branch, as it is not strong enough to bear their weight. In this manner they secure their nests.

There are also other large birds in these countries, called by us Guinea hens, 'galine de faraon', which are brought from the Levant. There are many of these hens, and geese, which do not resemble ours, having different feathers, and also divers other beautiful birds, large and small, of other species than ours.

CHAPTER XXXI

A MARKET, *and the* PEOPLE *who went thither*

Since it fell to me to spend many days on shore, I decided to go to see a market, or fair, at no great distance [from the spot where I was lodged]. This was held in a field, on Mondays and Fridays, and I went two or three times to it. Men and women came to it from the neighbourhood country within a distance of four or five miles, for those who dwelt farther off attended other markets. In this market I perceived quite clearly that these people are exceedingly poor, judging from the wares they brought for sale—that is, cotton, but not in large quantities, cotton thread and cloth, vegetables, oil and millet, wooden bowls, palm leaf mats, and all the other articles they use in their daily life. Men as well as women came to sell, some of the men offering their weapons, and others a little gold, but not in any quantity.

[1] Weaver birds, not parrots. See Appendix.

They sold everything, item by item, by barter, and not for money, for they have none. They do not use money of any kind, but barter only, one thing for another, two for one, three for two.

These negroes, men and women, crowded to see me as though I were a marvel. It seemed to be a new experience to them to see Christians, whom they had not previously seen. They marvelled no less at my clothing than at my white skin. My clothes were after the Spanish fashion, a doublet of black damask, with a short cloak of grey wool over it. They examined the woollen cloth, which was new to them, and the doublet with much amazement: some touched my hands and limbs, and rubbed me with their spittle to discover whether my whiteness was dye or flesh. Finding that it was flesh they were astounded.

To this market I went to see further strange sights, and also to find out whether any came thither with gold for sale, but altogether, as I have said, there was little to be found.

CHAPTER XXXII

Of the HORSES *they buy, their* FODDER *and the* HORSE CHARMERS

Horses are highly prized in this country of the Blacks, because they are to be had only with great difficulty, for they are brought from our Barbary by the Arabs and Azanaghi and cannot withstand the great heat, growing so fat that the greater part of them die from a malady in which they are unable to make water and so burst. Their forage in these parts is the leaves of the beans which are left in the fields after harvest. These are chopped very fine and dried like hay, and given to the horses for fodder in place of corn. They also give them millet, upon which they fatten greatly. A horse with its trappings is sold for from nine to fourteen negro slaves, according to the condition and breeding of the horse. [When a chief buys a horse] he sends for his horse-charmers, who have a great fire of certain herbs lighted after their fashion, which makes a great smoke. Into this they lead the horse by the bridle, muttering their spells. Then they have it rubbed all over with an ointment, and keep it for fifteen to twenty days without anyone seeing it. Then they fasten to its neck charms compressed into a small space and covered with red leather[1]. They believe that with these they are safer in battle.

CHAPTER XXXIII

The WOMEN *who* DANCE *by night*

The women of this country are very pleasant and light-hearted, ready to sing and to dance, especially the young girls. They dance, however, only at night by the light of the moon. Their dances are very different from ours.

These negroes marvelled greatly at many of our possessions, particularly at our cross-bows, and, above all, our mortars. Some came to the ship, and I had them shown the firing of a mortar, the noise of which frightened them exceedingly. I then told them that a mortar would slay more than a hundred men at one shot, at which they were astonished, saying that it was an invention of the devil's. The sound of one of our country pipes, which I had played by one of my sailors, also caused wonderment. Seeing that it was decked out with trappings and ribbons at the head, they concluded that it was a living animal that sang thus in different voices, and were much pleased with it. Perceiving that they were misled, I told them that it was an instrument, and placed it, deflated, in their hands. Where-

---

[1] These charms, usually containing texts from the Quran, are still very commonly worn on horses. Richard Jobson (*Golden Trade*, p. 63) gives a very similar description of them: 'The Gregories bee things of great esteeme amongst them, for the most part they are made of leather of severall fashions, wounderous neatly, they are hollow, and within them is placed, and sowed up close, certaine writings, or spels, which they receive from their Mary-buckes, whereof they conceive such a religious respect, that they do confidently beleeve no hurt can betide them, whilst these Gregories are about them, and it seemes to encrease their superstition.'

upon, recognising that it was made by hand, they said that it was a divine instrument, made by God with his own hands, for it sounded so sweetly with so many different voices. They said they had never heard anything sweeter.

They were also struck with admiration by the construction of our ship, and by her equipment—mast, sails, rigging, and anchors. They were of opinion that the portholes in the bows of ships were really eyes by which the ships saw whither they were going over the sea. They said we must be great wizards, almost the equal of the devil, for men that journey by land have difficulty in knowing the way from place to place, while we journeyed by sea, and, as they were given to understand, remained out of sight of land for many days, yet knew which direction to take, a thing only possible through the power of the devil. This appeared so to them because they do not understand the art of navigation [the compass, or the chart].

They also marvelled much on seeing a candle burning in a candlestick, for here they do not know how to make any other light than that of a fire. To them the sight of the candle, never seen before, was beautiful and miraculous. As, in this country, honey is found, they suck the honey from the comb, and throw away the wax. Having bought a little honeycomb, I showed them how to extract the honey from the wax, and then asked whether they knew what it was that remained. They replied that it was good for nothing. In their presence, therefore, I had some candles made, and lighted. On seeing this, they showed much wonderment, exclaiming that we Christians had knowledge of everything.

In this country they have no musical instruments of any kind, save two: the one is a large Moorish 'tanbuchi', which we style a big drum; the other is after the fashion of a viol; but it has, however, two strings only, and is played with the fingers, so that it is a simple rough affair and of no account.

CHAPTER XXXIV

How ANTONIOTTO *and* ALOUISE *sailed in company towards* CAPO VERDE

As I have said, I had occasion to remain several days in the country of the lord Budomel, that is, to see and to understand everything, and to trade. When I had despatched my business, and had acquired a certain number of slaves, I decided to continue beyond Capo Verde, to discover new lands, and to make good my venture: for before my departure from Portugal, I had understood from the Lord Infante, (as a person who from time to time had reflected upon the question of these countries of the Blacks which the caravel had reached, and upon other information,) that not very far beyond this first kingdom of Senega there was another kingdom called Gambra[1], where the negroes who had been carried off to Spain, or rather to Portugal, said, there was gold in large quantities: and that the Christians who should go thither would become rich. Hence I, moved by the desire to find this gold, and to see further novelties, despatched my business with Budomel, and returned to the caravel.

After making sail at once to leave this coast, I beheld one morning two vessels in the offing. Having sighted each other, we drew near to hail them, knowing that they could only be Christians. We learnt that one ship belonged to Antonioto Uso da mār[2], a Genoese, the other to certain squires of the said Lord Infante, who were sailing in company on a voyage of discovery and trade beyond the said Capo Verde. When I found that their plans were similar to mine, I joined them. With this object in common, our three caravels set a course for this Cape, maintaining a southerly direction along the coast, always within sight of land. Accordingly the next day,

[1] Gambia.
[2] For Usodimare's letter to his creditors, see Magnaghi, *op. cit.*, p. 28.

CAPO VERDE 53

aided by a favourable wind we sighted the Cape, which is distant about forty[1] of our [Italian] miles from my place of departure.

CHAPTER XXXV

CAPO VERDE; *the meaning of the name, and the customs of its inhabitants*

This Capo Verde is so called because the first to discover it (who were Portuguese) about a year before[2] I was in these parts found it all green with great trees, which remained in leaf throughout the year. For this reason they gave it the name of Capo Verde: just as Capo Bianco, of which we have already spoken, was found entirely sandy and white and was therefore called 'Capo bianco'. This Capo Verde is very beautiful and lofty: on the point there are two hillocks[3]. It runs far into the sea, and on the cape and in its vicinity there are many dwellings of negro peasants, huts of straw, close to the sea, and visible to those who pass. These negroes belong to the said Kingdom of Senega.

[Off the cape there are dry patches which extend about half a mile out to sea.] Off it we found three small islands, not very far from the land, uninhabited and covered with tall green trees. Being in need of water, we anchored off one of them[4], which appeared the largest and most fruitful, to ascertain if any springs were to be found there. On landing we found no water, except in one spot where there was a little water, but which was of no use to us. We found many nests on the island, and eggs

[1] From this it appears that he did not travel overland to rejoin his ship on the Senegal.
[2] Cape Verde was reached by Nuno Tristão in 1443, twelve years before Cadamosto's visit.
[3] 'The peninsula of Cape Verde is composed of moderately high land rising gradually to the Paps, two conical hills above Cape Verde itself....The Paps are quite distinct and are visible a good sixteen miles; during the rainy season they are covered with stunted vegetation' (*Africa Pilot*, Pt. I, p. 195).
[4] Madeleine Islands: once covered with vegetation, though now desert. Azurara, p. 324.

of various birds [un]known to us. While we remained here we all fished with lines and large hooks and caught a great number of fish: among them shell fish and very large mature dories, weighing from twelve to fifteen pounds each. This was in the month of June.

Thence, the following day, we continued to sail on our voyage, always within sight of land. Beyond Capo Verde there is a gulf inland[1]. All the coast is low, covered with very fine, tall, green trees, which never shed their leaves throughout the year, [that is they never wither, as do ours,] for new leaves appear before the old fall. These trees come right down to within a bowshot of the beach, so that it appears as though they flourished in the sea—a very beautiful sight[2]. In my opinion, who have sailed to many places in the Levant and in the west, I have never seen a more beautiful coast than this appeared to me—watered by many rivers, and streams of little account, since large vessels cannot enter them.

Beyond this small gulf, the whole coast is inhabited by two races, the one called Barbazini[3], the other, Sereri[4]: they are negroes, but not subject to the King of Senega. They have neither king nor lord of their own, but they nevertheless honour one more than another according to their birth and estate. They will not recognize any lord among them, lest he should carry off their wives and children and sell them into slavery, as is done by the kings and lords of all the other lands of the negroes. They are exceedingly idolatrous, have no laws, and are the cruellest of men. They use bows and poisoned arrows[5]; should an arrow touch the bare flesh and draw blood, the wound is at once fatal. They are very black, and well built. Their country is closely wooded, full of lakes and streams: from these they derive great security, because it is impossible to penetrate thither except by narrow tracks. They have therefore no fear of any

---

[1] Gorée Bay.　　　　　　　　[2] Mangrove swamps.
[3] The Barbacenes.
[4] The Serer: closely allied to the Jolof. 'The Serer have the reputation of being extraordinarily tall, but actual measurements hardly bear this out. They are less black than the Wolof, but have coarser features' (Seligman, C. G., *op. cit.*, p. 59).
[5] The most common arrow poison in this part of Africa is strophanthus.

neighbouring lord: it has frequently happened that kings of Senega in the past have sought to wage war to subdue them, and each time they have been roughly handled by these two races[, either through the poisoned arrows they use, or from the natural strength of the country].

Running with the wind along this coast, still voyaging southwards, we discovered the mouth of a river, perhaps a bowshot wide, and of no great depth. To this river we gave the name of Rio di Barbazini[1], and thus it is named on the 'carta da navigar' of this country made by me[2]. It is distant sixty miles from the Capo Verde. We always navigated this coast and beyond by day, anchoring each evening at a deserted spot in ten or twelve *passa*[3] of water, and four or five miles from the shore. At dawn we made sail, always stationing one man aloft and two men in the bows of the caravel to watch for breakers which would disclose the presence of shoals.

Sailing thus we reached the mouth of another large river, which appeared to be no smaller than the Rio de Senega. When we saw this fine river[4], and the beautiful country, we cast anchor, and debated whether we should send ashore one of our interpreters—for each of our ships had negro interpreters on board brought from Portugal, who had been sold by the lords of Senega to the first Portuguese to discover this land of the Blacks.

These slaves had been made Christians in Portugal, and knew Spanish well: we had had them from their owners on the understanding that for the hire and pay of each we would give one slave to be chosen from all our captives. Each interpreter, also, who secured four slaves for his master was to be given his freedom. Lots were drawn to decide whose interpreter was to go ashore, and it fell to the Genoese. Whereupon having fitted out his boat, he sent off his man, with orders that the boat was not to be run ashore except in as far as was necessary to land the slave. He was instructed to ascertain the condition of the

[1] The Joal.
[2] This reference to a chart made on the voyage occurs in the Marciana MS. (*Ital.* VI. 485) and in the first printed edition, but is omitted by Ramusio.
[3] Two *passa* equal approximately one fathom.
[4] The estuary of the Solum and Jumbas rivers.

country, to whom it was subject, and whether gold and other objects of use to us were to be obtained there. Accordingly, when he had landed, and the boat had withdrawn a short distance, he suddenly encountered a great number of negroes who, having observed the ships approaching, had lain in wait with bows and arrows and other weapons to accost any of our men who might land. They conversed for a short while; what he said to them we do not know, but they began furiously to strike at him with their short Moorish swords, and quickly put him to death, those in the boat being unable to succour him.

When we were informed by our men of this, we were left stupified, realising that they must be very cruel men to do such a thing to a negro of their own race, and that they might reasonably be expected to treat us much worse. On this account, we set sail, still holding our southerly course within sight of the shore, which appeared to us continually more beautiful, more thickly covered with green trees, and always low, until at length we reached the mouth of the river of Gambra[1]. Perceiving it was very wide, not less than three or four miles in the narrowest reach, so that our ships could enter in safety, we decided to lie there with the intention of ascertaining on the following day whether this was the country of Gambra, which we so greatly desired to find.

CHAPTER XXXVI

A GREAT RIVER, NAVIGATED *by* CANOES

On entering the river, which at its mouth is no less than six to eight miles wide, we concluded that the country must be Gambra, so ardently sought by us, and that along the river it would not be surprising to find some favourable spot where we might easily chance upon profitable business in gold or in other valuable commodities. The wind on the next day being very favourable, we sent the small caravel ahead, well manned by men from our vessels, with orders that, as the caravel was small

[1] The Gambia; Nuno Tristão had passed its mouth on his voyage of 1446.

THE COUNTRY OF GAMBRA

and drew little water, they were to advance as far as possible, and if they should encounter any banks in the estuary, to take soundings. If they should find sufficient depth of water for our vessels, they were to withdraw and, after signalling, to us to anchor.

These instructions were followed out by her, for, finding about two fathoms of water at the mouth, she anchored according to orders. When the caravel had come out, it was decided to send our boat, armed, in company with hers, farther up the river, since they were small. Their instructions were that, should the negroes come in their boats to assail them, they must return at once to the ship without attempting to fight. This was because we had come thither to trade in the country peacefully and with their approval, which would be more fittingly accomplished by tact than by force. The boats having proceeded ahead, they took soundings in many places and finding nowhere less than eight fathoms of water, continued for two miles. The banks of the river proved to be very beautiful, bordered with high green trees; since the river made many turns higher up, it appeared unnecessary to them to proceed farther.

On the return, there issued from the mouth of a stream which flows into this great river, three canoes (we call them *zopoli*) which, from what I observed later, are all made of a single portion of a large tree hollowed out, fashioned like the little boats which are towed behind our ships. When our boats saw these canoes, being doubtful whether they might not have come to do them injury, and having been warned by other negroes that in this country of Gambra all the bowmen used poisoned arrows, they took to their oars in obedience to their instructions [and to avoid a quarrel], although they were sufficiently numerous to defend themselves, and returned with all possible speed to the ship. They did not return so rapidly however but that the canoes were close behind, within a bowshot of them, when they reached the ship, for they are very swift. When our men had boarded their ship, they began to gesticulate and to make signs to the canoes to draw near. These slowed down, and approached no nearer. There were about twenty-five to thirty negroes in each; these remained for a while gazing upon a thing which neither they nor their fathers had ever seen before, that is ships and white men, without showing any wish to parley, despite all that was said to them, and then went about their own affairs. And so that day passed without further incident.

CHAPTER XXXVII

How the BLACKS ATTACKED *the* SHIPS *in the* RIVER

The following morning, at about the third hour, we on the other two ships, made sail with a favourable wind and tide to seek our consort and in God's name to enter the river, hoping that in the country farther upstream we might find more civilized people than those we had seen in the canoes. Having joined our consort, she made sail in company and we began to enter the river: the small caravel led the way over the shallows, we following one behind the other.

Having sailed about four miles upstream, we suddenly perceived several canoes coming up behind us (I do not know from whence they came) as fast as they were able. Seeing this, we turned upon them, and being dubious of their poisoned arrows of which we had been informed we protected our ships as best we could, and stood to arms at our stations, although we were poorly equipped. In a short time they reached us. I, being in the leading ship, split the canoes into two sections, and thrust into the midst of them: on counting the canoes, we found they numbered seventeen, of the size of considerable boats. Checking their course and lifting up their oars, their crews lay gazing as upon a marvel. We estimated on examination that there might be about one hundred and fifty at the most; they appeared very well-built, exceedingly black, and all clothed in white cotton shirts: some of them wore small white caps on their heads, very like the German style, except that on each side they had a white wing with a feather in the middle of the cap, as though to distinguish the fighting men. A negro stood in the prow of each canoe, with a round shield, apparently of leather, on his arm. They made no movement towards us, nor we to them; then they perceived the other two vessels coming up behind me, and advanced towards them. On reaching them, without any other salute, they all threw down their oars, and began to shoot off their arrows.

Our ships, seeing the attack, at once discharged four bombards: hearing these, amazed and confounded by the roar, they threw down their bows, and gazing some here, some there, stood in astonishment at the sight of the shots falling into the river about them. After watching thus for a considerable while, and seeing no more they overcame their fear [of the thunder claps after many shots had been fired], and taking up their bows, began afresh to shoot with much ardour, approaching to within a stone's throw of the ships. The sailors began to discharge their cross-bows at them: the first to do so was a bastard son of the Genoese, who hit a negro in the breast so that he immediately fell dead in the canoe. His companions perceiving this pulled out the arrow and examined it closely, in astonishment at such a weapon: but this did not restrain them from shooting vigorously at the ships, the crews of which replied in like fashion so that in a short space a great number of negroes were wounded. By the grace of God, however, not one of the Christians was hit.

When they saw the wounded and dead, all the canoes with one accord made for the stern of the small caravel, where a stiff fight was waged, for her crew were few and ill armed. Seeing this I made sail for the small vessel and towed her between our two larger ships amidst a discharge of bombards and cross-bows. At this, the negroes drew off: we, lashing our three ships together by chains, dropped anchor, which, [as the water was calm] held all three.

We then attempted to parley with the negroes.

CHAPTER XXXVIII

The COUNCIL *which was held on the river of* GAMBRA

After much gesticulating and shouting by our interpreters one of the canoes returned within bowshot. We asked of those in it the reason for their attack upon us notwithstanding that we were men of peace, and traders in merchandize, saying that we had peaceful and friendly relations with the negroes of the Kingdom of Senega, and that we wished to be on similar terms with them, if they were willing: further, that we had come from a distant land to offer fitting gifts to their king and lord on behalf of our king of Portugal, who desired peace and friendship with them. We besought them to tell us in what country we were, what lord ruled over it, and the name of the river, and told them they might come in peace and confidence to take our wares, for we were content that they should have as much or as little as they pleased.

They replied that they had had news of our coming and of our trade with the negroes of Senega, who, if they sought our friendship, could not but be bad men, for they firmly believed that we Christians ate human flesh, and that we only bought negroes to eat them: that for their part they did not want our friendship on any terms, but sought to slaughter us all, and to make a gift of our possessions to their lord, who they said was three days distant. Theirs was the country of Gambra, and to the river, which was very large, they gave a name which I do not recall.

At this moment the wind freshened; realizing the ill will they bore us, we made sail towards them. They, anticipating this move, scattered in all directions for the land, and thus ended our engagement with them.

Thereupon, we debated [took the advice of our chief men, who formed the ship's council,] whether we should proceed farther up the river, if possible for at least one hundred miles, in the hope of finding better disposed peoples. But our sailors, who wished to return home and not to essay further dangers,

THE SOUTHERN CROSS

began with one accord to murmur, declaring that they would not consent to such a course, and that what had been done was sufficient for that voyage.

When we saw that this was their general desire we agreed to give way in order to avoid dissention, for they were pig-headed and obstinate men. Accordingly on the following day, we departed thence, shaping our course for Cape Verde to return, in God's name, to Spain.

CHAPTER XXXIX

The elevation of our NORTH STAR; *and the six stars opposite*

During the days we spent at the mouth of this river, we saw the pole star once only; it appeared very low down over the sea, therefore we could see it only when the weather was very clear. It appeared about a third of a lance above the horizon. We also had sight of six stars low down over the sea, clear, bright, and large. By the compass, they stood due south, in the following fashion[1]—

$$
\begin{array}{ccccc}
 & * & & & \\
* & & * & * & * \\
 & * & & &
\end{array}
$$

This we took to be the southern wain, though we did not see the principal star, for it would not have been possible to sight it unless we had lost the north star. In this place we found the night to be 13 [eleven and a half] hours, and the day 11 [twelve

---

[1] This figure is omitted from the printed text of the *Paesi*, 1507; the above diagram is from the Marciana MS. Ramusio's diagram differs slightly. As the Marciana MS. probably dates from *circa* 1470, it appears that as early as that date the constellation was given the form of a cross, although the first authentic drawing of it in that form was made by Mestre Joanes, who accompanied Cabral on his voyage of 1500. Fontoura da Costa does not doubt the possibility of Cadamosto having seen the Cross in June 1455: 'At 7 o'clock, at the beginning of the night, the α of the Cross would be about 8½° above the horizon, and the γ 12¼°, so that all the Cross would have been well visible to the south-west of the South Pole' (*A marinharia dos descobrimentos*, p. 113). Cadamosto's is the first recorded notice of the Southern Cross, though doubtless it had been seen by the Portuguese mariners who had preceded him.

and a half] hours, that is, in the first days of July, or more accurately on the second of the month.

This country is hot at all seasons of the year. It is true that there is some variation, and what they call a winter: thus beginning in the aforesaid month [of July] until the end of October it rains continuously almost every day from noon[1], in the following way: clouds rise continually over the land from the E.N.E., or from the E.S.E., with very heavy thunder, lightning and thunderbolts. Thus an excessive quantity of rain falls, and at this season the negroes begin to sow in the same manner as those of the kingdom of Senega. Their sustenance is entirely millet and vegetables, flesh and milk.

I understand that in the interior of this country, [on account of the great heat of the air] the rain which falls is warm. In the morning, when day breaks, there is no dawn at the rising of the sun, as in our parts, where between dawn and sunrise there is a short interval before the shadows of night disperse: the sun appears suddenly, though it is not light for the space of half an hour, as the sun is dull and, as it were, smoky on first rising. The cause of this appearance of the sun early in the morning, contrary to what happens in our country, cannot, I think, arise from any other circumstance than the extreme lowness of the land, devoid of mountains, and all my companions were of this opinion.

CHAPTER XL

How ANTONIOTTO the GENOESE and ALOUISE DA MOSTO discovered new islands

Little or nothing can be said of the features of this country of Gambra from what we were able to see and to learn on my first voyage—particularly from our own observations—for, as will have been understood, the people of the coast were so rude and

---

[1] The rains in this region last from June to October. At Gorée the mean annual range of temperature is 16° F., the coolest month being February. The tornadoes 'almost always travel from east to west, and are especially frequent at the beginning and end of the rains' (Kendrew, W. G., *Climates of the Continents*, p. 37).

savage that we were unable to have speech with them on land, or to treat about anything. We were then forced to return to Spain, without advancing further, for as I have already said, our sailors refused to proceed.

Thus, the following year[1], the same Genoese gentleman[2] and I, once again in agreement, fitted out two caravels with the purpose of exploring this river. Having informed the said Lord Infante (without whose licence we could not have gone) that we had reached this decision, he was greatly pleased, and had one of his caravels equipped to sail in our company. Rapidly furnished with all necessaries, we left the place called Lanchus[3], near Capo San Vincenzo in the beginning of the month of March[4], with a favourable wind, and laid our course for the Canaries, which we reached in a few days. The season being favourable, we did not think it worth while to touch at these islands, but sailed steadily southwards on our voyage. And, with the assistance of a current [the waters] that flowed strongly to the south-west, we sped along very rapidly. Finally we reached Capo Bianco: Having picked it up, we stood off to sea a little. The following night there arose a storm from the south-west with a strong wind: upon which, in order not to turn back we held our course towards west-north-west, to the best of my belief, to sail as near the wind as possible[5], for two nights and three days.

The third day we sighted land, everyone shouting 'Land, land', and marvelling exceedingly, because we were not aware of any land in these parts. Two men ordered aloft descried two large islands[6]. When we heard of this we gave thanks to our Lord God, who had led us to see new things, for we knew well that no such islands as these had been reported in Spain. [We judged that these might be inhabited.] To ascertain further and to test our fortune we made for one of them, and in a short time

---

[1] That is to say 1456. [2] I.e. Usodimare. [3] Lagos.
[4] The printed text of the *Paesi* has 'May': I have followed the Marciana MS. here, as March appears more probable.
[5] Literally 'in order to parry and go by the side of the weather [or storm]'.
[6] The Cape Verde Islands; for a discussion of priority in their discovery see Introduction, p. xxxvi.

drew near it[1]. As, on arrival, it seemed large, we coasted for a short distance within sight of the shore until we reached a place which appeared to be a good anchorage: there we dropped anchor, and as the weather was calm, hoisted out the boat: this, well manned, was sent away to the land, to ascertain if there were any people on the island. The boat put off, and reconnoitred much, but found no tracks nor signs from which it might be concluded that there were inhabitants. When they had reported this to me, in order to satisfy myself completely, I sent the following morning ten men, well equipped with weapons and cross-bows, with orders to climb a mountainous and high part of the island[2], to see whether they could find anything, or catch sight of other islands. This they did, without ascertaining anything except that the island was uninhabited, and that there was an exceeding number of doves, which allowed themselves to be taken by hand, being unused to man. They brought back to the caravels many which they had taken with staves and clubs. From the farther shore they had sighted three other large islands, which we had not seen. One of these lay to the leeward towards the north: the other two were in line to the south, on our course, all being in sight of each other. They could also see in the western quarter, far out to sea, what appeared to be other islands, but they could not make these out clearly on account of the distance. I did not choose to go to these, as I wished to lose no time in continuing my voyage, and as I judged that they would be uninhabited and wild like the others: but afterwards when others were drawn thither by the news of the four islands I had found, these others were sought out: they proved to number ten, both large and small. They were uninhabited and nothing was found there but doves, strange species of birds, and large shoals of fishes.

But to return to my narrative: leaving this island and resuming our course, we sighted the other two islands. Then, sailing along the coast of one of them[3], which appeared well-

[1] The island afterwards called Bonavista (see below, p. 65), and by the Portuguese, S. Cristovão.

[2] The highest point on the island is the Pic Estancia, 387 m.

[3] Called by Cadamosto 'San Iacomo', and by the Portuguese Santiago (see below, p. 101). It is the largest island of the archipelago.

ISOLA DE BONAVISTA

wooded[1], we discovered the mouth of a river issuing from it. Judging that there the water would be good, we anchored to replenish our supplies. Some of my men, having landed, went along by the bank to the mouth of this river, where they found small quantities of very white, pure salt, some of which they brought back to the ship. Of this we took as much as we required: also, finding the water very good, we replenished our supply. I should also explain that we found here large numbers of turtles (what we call *gajandre*), some of which we caught[2]. Their shells are larger than good-sized bucklers. These the sailors killed and added to our victuals, for they said that on other occasions in the Golfo de Argin, where they are also found but not of such size, they had eaten them. I may also say that, to test everything, I ate some, and found them good, not unlike the white flesh of veal, so good was their smell and taste. We accordingly salted many of them, because they made good supplies for the voyage. I also ate on my first voyage the flesh of the elephant, which I did not consider very good. We also fished at the mouth of this river and higher up, where we found so great a quantity of fish that it is incredible to record. Many species we had never before seen were large and of fine flavour. The river was sufficiently deep for a vessel of 150 tons burthen to enter with ease—a good bowshot in width. We stayed there for two days to rest, laying in the aforementioned supplies, including many doves, of which we killed immense quantities. It is to be noted that to the first island upon which we landed we gave the name 'Isola de Bonauista', as it was the first land sighted in these parts: to this other island, which appeared to be the largest of the four, we gave the name 'Isola de San Iacomo', because it was on San Philippo Iacomo's Day that we came to anchor there[3].

[1] This is correct; the forests were practically all destroyed by the early colonists, but small remnants of the original covering are still to be seen. A recent visitor mentions *Faidherbia albida*, a member of the mimosa family, and *Ficus gnaphalocarpa*, a wild fig, as being most characteristic of this (Chevalier, A., *Iles du Cap Verd*, p. 757).

[2] Turtles are numerous on the island, the commonest marine turtle being *Testudo midas*. Large quantities were salted down and sent to feed slaves at S. Vicente in the seventeenth century (Chevalier, A., *op. cit.*, p. 788).

[3] The Feast of St Philip and St James is on May 1st. There is no explanation in the text, however, why they had spent over a month in sailing from the Canary Islands.

CHAPTER XLI

'LE DO PALME', *and the navigation of the river of* GAMBRA

Having done what I have written above, we left these four islands, setting a course for Capo Verde; whither in a few days, with God's help, we came close in shore, in sight of land at a place called 'Le do palme'[1] between Capo Verde and Rio de Senega. As we had good knowledge of the land, we continued towards the Cape, and on the following morning passed it. We sailed on until we arrived for the second time at the aforementioned river of Gambra, which we entered at once, and without coming into further contact with the Blacks or their canoes, we sailed up river by day, with the lead always ready. The canoes of the negroes, a few of which we came across, kept along the river banks[, not daring to accost us]. About ten miles upstream we came upon an islet, similar to a *polesine*[2], formed by the river[3]. When we had cast anchor by it, on a Sunday, one of our sailors, who had been prostrated many days by fever, departed this life, and though his death depressed us all, nevertheless, wishing to do that which would be pleasing to God, we buried him on this island. His name was Andrea; for which reason we decided that the island should in future be called 'Isola di Sancto Andrea', and thus it has always been known.

Leaving the island and sailing always upstream, we were followed at a distance by some canoes of the negroes. Attracting their attention, our interpreters called to them, displaying silken

[1] These were a notable landmark near the mouth of the Senegal river. Azurara, referring to Lançarote's expedition, says: 'Now these caravels, having passed by the land of Sahara, as hath been said, came in sight of the two palm trees that Dinis Diaz had met with before, by which they understood that they were at the beginning of the land of the Negroes' (Hakluyt Soc., C, p. 176). The text is in error in implying that they were south of the Senegal.

[2] A marshy area near a river, derived from the regional name for the lower Po valley.

[3] If the ten miles are taken from the mouth of the estuary, this may be the islet off Dog Island Point.

THE COUNTRY OF GAMBRA

stuffs and other articles, and gave them to understand that they might with safety draw near, that we would give them these garments, and that they need have no fear, for we were humane and well disposed men. Little by little the negroes drew nearer, gaining confidence in us, until at last they drew alongside my caravel, and one of them who could understand my interpreter boarded the ship. He marvelled greatly at her, and at the method of navigating by means of sails, for they knew no method except by rowing their canoes with oars, and considered no other way possible. He was overcome with astonishment at the sight of us white men, and marvelled no less at our clothing, so different to his—principally in that most of them went naked, or, if clothed, in a white cotton singlet. We made much of the negro, giving him many trifles of little value, with which he was exceedingly pleased, and asked many questions.

At last I ascertained that this was the land of Gambra, and that the principal lord was Farosangoli[1], who, he said, dwelt far from the river inland towards the south-south-east, according to his directions, at nine or ten days' journey. This Farosangoli was subject to the Emperor of Melli, the great Emperor of the Blacks, but nevertheless, there were many lesser lords who dwelt near the river, some on one bank, some on the other: he offered, if I were willing, to bring me to one of them, called Batimaussa[2], and to treat with him to enter into friendship with us, since it seemed to him that we were well disposed persons. This offer was very acceptable to me: so, taking him on board and treating him hospitably, we sailed up the river until we reached the place of the said Batimaussa, which according to our estimate was about sixty miles and more from the river mouth.

[1] According to Delafosse (*Haut-Sénégal-Niger*, II, p. 21), *faran* was a Songhai name for a chief.
[2] *Mansa* in the Mandingo language means king or chief, cf. Mansa Musa.

CHAPTER XLII

How the LORD BATIMAUSSA *offered friendship to us*

Note that while ascending this stream we were going eastwards and that at the spot where we dropped anchor the stream was much narrower than at the mouth, being in our judgment no more than a mile in width. This river has many branches which join together.

Arrived at this place, we decided to send one of our interpreters with the negro into the presence of this lord Batimaussa, with a present for him, a *zuba* [*alzimba*] of Moorish silk—as we should say a surcoat—which was quite fine and made in the land of the Moors. He was to say that we had come by command of our lord King of Christian Portugal, to establish firm friendship with him, and to inform him that if he had need of the products of our country, our King would send them to him each year, and many other messages.

The interpreter went with the negro to this lord, and, in brief, we treated so with him that when we parted from him we had not only secured his friendship, but had bartered many articles, for which we received in exchange negro slaves and a certain quantity of gold, but not of much account with respect to what we had anticipated, because the rumour of it had been much greater in the reports of negroes of Senega: indeed we found very little by our standards, but by those of these very poor people it was considerable. Gold is much prized among them, in my opinion, more than by us, for they regard it as very precious: nevertheless they traded it cheaply, taking in exchange articles of little value in our eyes.

We remained here about fifteen days[1], and in this time many negroes dwelling on a part of the border of this river came to our ships—some to gaze upon a sight so strange to them: others to sell some trifle of theirs, or little rings of gold. The articles

[1] Thus in the '*Paesi*'; the Marciana MS. has 'about two days'; Ramusio 'about eleven days'.

they brought were cotton cloth and thread, cotton cloths woven in their fashion, some white, others variegated, white and blue striped, or red, blue, and white, excellently made. They also brought many apes, and baboons of various species large and small, of which there are very large numbers in these parts. These they bartered for objects of little worth, giving ten *marchetti* for the value of one. They also brought for sale civet, and the skins of cats from which civet is obtained. They gave an ounce of civet in exchange for an article not worth forty or fifty *marchetti*, not that they sold by weight, but as I say by estimation. Others brought fruits of various kinds, among them many small wild dates, which they eat. Many of our sailors ate them, and found them of a different flavour to ours, but I had no desire to eat them, for fear of dysentery. In this way we had each day fresh people of various tongues down at the ships. They are constantly journeying from place to place up and down the river in their canoes, with women and men, as with us our boats do on the floods.

But all their navigation is by means of oars: they row standing up, so many on each side. They always have one extra rower at the stern, who rows now on one side, now on the other, to keep the boat straight. They do not use rowlocks for their oars, but hold them steady with their hands. The oars are fashioned thus: they have a shaft, like a short lance, a yard and a half in length: at the head of this shaft they nail, or rather bind after their fashion, a round disc: with this style of oar they row their boats exceedingly swiftly along the shore from point to point. There are many mouths of streams which they enter and leave in safety, but they do not commonly venture far outside their own country, for they are not safe from one district to the next from being taken by the Blacks and sold into slavery.

At the end of a certain time[1] we resolved to depart thence, and to proceed to the mouth of the river, because many of our men began to suffer from a high fever, sharp and continuous: thus we left suddenly.

[1] 'Two days' in Marciana MS.

CHAPTER XLIII

WILD ELEPHANTS *and the method of hunting them*

Of the things to be recounted of this country, from our own observations, and from the information we gathered during the short time we were there, we shall deal first with their faith. It is, in general, idolatry in various forms: great credence is placed in spells, and other diabolical methods with which they are acquainted, but all recognize one God, and some of them hold the tenets of Muhammad. The latter are men who frequent other countries, not remaining tied to their homes, for the peasantry know nothing of such things.

In their way of life they conduct themselves in almost all respects similarly to the negroes of the kingdom of Senega; they eat the same foods, except that they have more varieties of rice than grow in the country of Senega: also they eat the flesh of dogs, which I have never heard is eaten elsewhere. Their garments are of cotton, whereas almost all the negroes of Senega go naked. They are clothed because of the great abundance of cotton. The women are also clothed in a similar style, except that they delight in their youth to work designs upon their flesh with the point of a needle, either on their breasts, arms, or necks. These appear like those designs of silk that are often made on handkerchiefs: they are made with fire, so that they never disappear.

This region is very hot, and the more one advances towards the south, the hotter these countries become. On this river particularly it is very much hotter than on the sea, because it is covered with numerous and very large trees which are everywhere throughout the country. Concerning the size of these, I may say that, at a spring near the river bank from which we drew water, there was a very great and broad tree; its height, however, was not in proportion to its size, for while we judged it to be about 20 paces high we found the girth by measurement to be about seventeen paces round the foot. It was hollow in many places, and its branches were very large so that they threw a deep shade around[1]. There are to be found even larger trees, so that from such trees it may be concluded that the nature of the country is good, and fertile, being bathed with many waters.

There are large numbers of elephants in this country. I have seen three wild ones, for they do not know how to domesticate them as in other parts of the world. When we caught sight of these three elephants emerging from the forest, the ship was lying in midstream: some of us jumped into the boat to go to them for they were some distance off, but when the animals saw us approaching, they returned to the forest. Later I saw a small one, dead: for, to satisfy me, a negro lord named Gnumimenssa[2], dwelling near the mouth of this river of Gambra, set out to hunt it with many negroes, following for two days before they killed it. They go hunting on foot, carrying no other weapons for the attack save the *azagaie* which I have described above and bows, all their weapons being poisoned. You should know that they seek the elephants in woods, for these prefer swampy places, where for the most part they resort, like swine. The negroes place themselves behind the trees, and wound the elephant with arrows or poisoned spears. They advance scrambling and jumping from tree to tree, so that before the elephant, which is an unwieldy animal, can escape, it is wounded in many places without being able to defend itself: I may say here that in the open, with no trees near, no man would dare to face one, for no man can run so fast that the elephant, without breaking out of its ordinary pace, cannot overtake him; for considering his size his pace is very rapid: if it happens by a mischance that an elephant pursues a man in the open and overtakes him he attacks him with nothing but his great trunk, which is somewhat like that of a pig, except that the pig's snout is not mobile, as is the elephant's. It resembles a large tough lip, which, unlike the pig, he is able to twist, extend and shorten at will; winding this trunk around the man he hurls him so far into the air that often he is dead before he falls to the ground. This was told me by many negroes. But the elephant is not however so ferocious an animal that it will attack men without first being annoyed.

[1] The Baobab.
[2] King, or chief, Gnumi.

I saw this small elephant lying dead on the ground. Its tusks were no more than three spans in length. Of these three spans, one was embedded in the jaw, so that the tusks were actually only two spans in length. For this reason they said it was a young animal, since some have tusks from ten to twelve spans long: but small though it was, we judged that its carcass equalled those of five or six of our bulls.

I was given this elephant by the chief, that is, I was allowed to take whatever portions of it I wished, and the remainder was given to the hunters for food; gathering from this that its flesh is eaten by the negroes, I had a portion cut off, which, roasted and broiled, I ate on board ship, to establish this thoroughly, and to be able to say that I had eaten of the flesh of an animal which had never been previously eaten by any of my countrymen. The flesh, actually, is not very good, seeming tough and insipid to me. I also brought away one of its feet and a portion of its trunk to the ship, and also many of the hairs from its body, a span and a half in length or more, and very thick. These, with a portion of the flesh [which I had salted down], I presented later in Spain to the lord don Heurich, who received them as a handsome gift, being the first that he had had from the country discovered through his energy.

CHAPTER XLIV

Of the feet and limbs of the elephant: and of the river-horse

I wish it to be understood that the foot of the elephant is round, almost like the foot of the horse, but the foot has not a hoof like the horse, but is all a black, thick, callosity: around this there are five claws, level with the ground, round, and little larger than a *grossone*. The foot of this little elephant was not so small but that it was a span and a half in width, across the sole in every direction, for, as I have said, it is quite round, like a platter.

THE HIPPOPOTAMUS

I was also given by the same chief another elephant's foot which I measured several times across the sole in the presence of many people, and found it to be three spans and an inch in breadth in either direction. I also presented this to the lord Infante, with a tusk twelve spans long, which with the said foot he ordered to be given to the lady Duchess of Bergogna[1] as a worthy present. Do not believe that the elephant cannot bend its knees, as I have heard said at times; on the contrary it moves, bends, and rises like any other animal.

Also in this river of Gambra, and in many other rivers in the country, in addition to the cockatrice and divers other animals, there is found an animal called a 'river horse'[2] (*pesse cavallo*); this animal is in nature something like the 'old man of the sea' (*vechio marin*), which lives now on the land, now in the water, and maintains itself in both elements. It is formed thus—its body, the size of a cow's, with short limbs, has cleft feet, and its head is shaped like a horse's, with two large tusks, as a wild boar has. These are very large, for I have seen some two spans in length, and at times longer. This animal frequently comes out of the water, and walks about on the banks like a quadruped: the like has not been found in any other parts to which we Christians have sailed, save these countries of the Blacks [as far as I can ascertain, unless, perchance, in the Nile.]

There are also to be found bats, or as we say *nottole*, of three spans and more, and many other birds of different kinds to ours, and especially innumerable parrots. This river is rich in fish, unlike ours in form and flavour, but none the less very good to eat.

CHAPTER XLV

The great river of CASAMANSA *and* CAPO ROSSO

As I have said above, we left [the port of Mansa, in] the country of the chief Batimansa, on account of illness among our men, and in a few days cleared the river. On issuing forth, it seemed

[1] The Duchess of Burgundy was Prince Henry's sister.
[2] The hippopotamus.

to us all, that, having sufficient victuals, it would be a praiseworthy action, since we were there, to cruise farther along this coast, for with three ships, we had sufficiently good company: accordingly, one day towards the third hour, we made sail with a favourable wind. As we were greatly engulfed in the mouth of the Rio de Gambra and the land towards the south-south-east ran well out to sea, appearing to form a cape, we set a westerly course to run well out to sea: and this land appeared low, covered with innumerable very beautiful, large green trees.

When we had stood out to sea a sufficient distance, we discovered that this could scarcely be called a cape, for beyond the point the coast-line was seen stretching away. Nevertheless we went well beyond the point, for around it the sea was breaking more than four miles off[1]. For this reason, we kept two men continually in the bows, and one at the masthead, to look out for rocks or other shoals. We navigated by day only, with as little sail as was necessary, kept strict watch, and anchored at night. The caravels sailed one after the other, as decided by lot each day, for each of us wished our consorts to lead the way, but all was decided by lot; one day it would fall to us, on another day to the others.

Navigating thus for two days along the coast, always in sight of land, we discovered on the third the mouth of a quite considerable river[2], which proved to be half a mile across. Continuing farther, we sighted towards evening a small gulf, which appeared almost like the mouth of a river. Accordingly, as it was late, we dropped anchor. The following morning making sail, and drawing inshore, we discovered the mouth of another large river[3]. It seemed, in my judgment, little smaller than the mouth of the river of Gambra: on either side of the river, great masses of very beautiful, tall, green trees were to be seen. Putting in, we cast anchor. After consulting together, we decided to arm two of our boats, and to send them ashore with our interpreters to gather information about the country—the name of the river, and of its chief. This was done; the boats put off

[1] Bald Cape and the reefs lying off it.
[2] Either the River Bliss or the River Suta. The 'small gulf' is perhaps the Oyster River.
[3] Kasamanze River.

and on their return stated that it was called 'rio de Casamanssa', that is to say, the river of a negro chief named Casamansa, who dwelt about thirty miles up the river, but that this chief was not now to be found there, having gone to war against another. After learning this, we left on the following day. Note that from the Rio di Gambra to this of Casamansa is about twenty-five leagues, that is, one hundred miles.

CHAPTER XLVI

The GREAT RIVERS *found on this coast, and the people*

Leaving the river of Casamansa we followed the coast until we reached a cape which we judged to be about twenty miles from the river. This cape is somewhat higher than the coast: as the face shows red, we gave to it the name Capo Rosso[1].

We continued along the coast beyond until we reached the mouth of a moderately sized river, in our estimate, a bow shot in width. This, to which we gave the name Rio de Sancta Anna[2], we did not choose to examine. Continuing upon our course beyond it, we came to another river on the coast, which did not appear smaller than that of Sancta Anna, and to this we gave the name Rio de Sancto Dominico[3]. We judged it by estimation to be fifty-five to sixty miles from Capo Rosso to this river.

Then, still sailing along this coast for a day, we came to the mouth of a very large river[4]. I say very large, for we all thought at first that it was a gulf: nevertheless very beautiful green trees were to be seen on the other part of the land to the south. We all estimated it to be at least twenty miles and upwards in breadth, so that we spent a considerable time in crossing this

[1] Cape Roxo: this had already been named by the Portuguese.
[2] Cacheu River.   [3] Mansoa River.
[4] The Rio Grande, now the River Jeba. Cadamosto's impression of its size is strongly emphasized on the charts of Benincasa.

estuary, that is from one side to the other: when we were at the farther side, we sighted islands out to sea.[1]

We then decided to obtain news of the country at this place, and at once anchored. The next morning came out to our ships two canoes, which were similar to those already described, and in truth were of great size; one was almost as long as one of our vessels, but not so high; and in it were more than thirty negroes; the other was smaller, having about sixteen men. On seeing them rowing out very fast, in the manner already described, and having doubts of them, we stood to arms, waiting to see what they were about. When they drew near, they raised a white cloth tied to an oar, as though requesting safe conduct. We replied in the same fashion; on seeing which, they came alongside, the larger to my caravel, and gazed upon us in great amazement to see white men. They also examined the shape of the vessel, the mast and the cross-yards, for they did not know what they were, nor their use. Then I, wishing to gain information of this people, caused my interpreters to speak with them, but none of them could understand what was said, nor could those on the other caravels: on realizing this, I was greatly disappointed, and at last they left without our having been able to understand them.

Reflecting that we were come to a new country of which we could not learn anything, we decided that to continue farther would be useless, for we judged that we should be continually encountering new dialects, and should not be able to achieve any good results. We accordingly decided to turn back. From the negroes in the canoes we obtained some gold rings in exchange for some trifles, buying and selling without speaking to them.

In this fashion we spent two days at the mouth of this great river. The north star appeared there very low in the sky. In this place we observed a most unusual phenomenon, which, as far as I have been able to learn, is not found in other parts to which Christians have sailed: the tides rose and fell, as they do at Venice and in all the West: but whereas in other places they flow for about six hours and ebb for another six, here they flow

[1] The Bissagos Islands.

RIO GRANDE 77

for four hours and ebb for eight[1]. So great is the force of the tidal stream when it begins to wax that it is almost incredible—three anchors at the bows with great difficulty could scarcely hold, and thus it was that the current made it necessary to set sail, and not without danger, for it was much stronger than the sails and the wind.

CHAPTER XLVII

How we DEPARTED THENCE, *as we could not* SPEAK *the* LANGUAGE

We left the estuary of this great river to return thence to Spain, and set our course at sea for those islands, which were about thirty miles from the mainland. We reached them, two large and several small ones[2]. The large ones are inhabited by negroes. The islands are very low, covered with very beautiful, large trees, tall and green. We had, however, no speech with the negroes, for they did not understand us, nor did we them. Departing thence, we proceeded towards our parts of Christendom, whither we sailed, so that God in his mercy brought us when it pleased Him safely to port.

[1] The *Africa Pilot*, with reference to the Jeba channel, warns vessels of the tidal streams and states that great caution is necessary. 'In the channel, both streams attain a rate of two to three knots; in its eastern part,...the ebb stream sets strongly towards Bissau island...off the mouth of the Orango channel the streams set for six hours each way' (Pt. I, p. 234). It is noteworthy that, referring to the Cacheu river, the *Pilot* states 'The ebb-stream usually runs eight hours, but the flood stream, which is often merely a lessening of the ebb stream, or slack water, rarely runs more than four hours' (p. 233). It is possible, therefore, that Cadamosto's remarks refer to his 'Rio de Sancta Anna', rather than to his 'Rio Grande'.

[2] The Bissagos Islands are shown thus on the Benincasa chart of 1468.

CHAPTER XLVIII

The DISCOVERERS *of new* COUNTRIES, *with the* NAMES *of the* LATTER

This is what I have seen and learnt during the time I was in these parts; but there have been others after me. Of most importance were the two armed caravels which the king of Portugal had sent thither after the death of the Infante Don Heurich[1]. Their commander was one Piero de Sinzia, a squire of this lord's, whom he commissioned to sail farther along this coast of the Blacks, and to discover new lands. With this captain went a Portuguese youth, a friend of mine, who had been thither with me as a notary. On the return of the caravels, I, Alouise da Mosto, was in Lachus[2], a place near Capo de San Vincenzo whither the said captain returned. My friend came ashore to my house, and gave me his observations point by point on the land they had discovered, the names they had given, the places at which they had stayed, all in due order commencing with the above mentioned Rio Grande[3]—whither we had previously been—as set forth below.

First my friend told me that they had been to the above mentioned large inhabited islands[4], upon one of which they had landed and spoken with the Blacks, but had been unable to understand them. They had also visited their dwellings some distance inland, miserable huts of straw. In some of these they found wooden idols, from which they concluded that these blacks were idolators, and worshippers of idols. Not being able to obtain anything nor to understand these people, they left them and continued their voyage along the coast until they reached the mouth of a great river, in his judgment three to four

[1] Prince Henry died in 1460. Cadamosto left Portugal in 1463.
[2] Lagos.
[3] The farthest point reached by Cadamosto: the River Jeba.
[4] The Bissagos Islands. According to the *Africa Pilot*, Pt. I, p. 234, 'The houses are circular, built of stones and mud, with low doors and no windows, rendering them dark and insufferably hot, though well thatched, with broad projecting eaves.' The inhabitants are still noted for their skill in fashioning wooden figures.

miles wide, which he estimated to be fifty miles along the coast beyond the mouth of the Rio Grande. He said that this river was called the Rio de Besegue[1], from the name of the chief that dwelt near its mouth.

Departing thence, and sailing on, they came to a cape which they called Capo de Verga[2]. All the coast between the Rio de Besegue and Capo de Verga is mountainous, though not of great height. The distance between them he estimated at one hundred and forty miles[3]. The mountains are covered with very beautiful, large, and tall trees, and appear very green from a great distance—a very pleasant sight to the eye.

CHAPTER XLIX

CAPO DE SAGRES, *with three others*; *and the name of the very great river*

They passed this Capo de Verga, and sailing along the coast for a distance of about eighty miles, discovered another cape, which in the opinion of all the sailors was the loftiest cape they had ever seen. In the middle of the land of this cape there is a high point, in the shape of a diamond. The whole is covered with very tall, green trees. They named it Capo Sagres[4] after a fort which

---

[1] The River Cassini.

[2] The Cape Verga of the Admiralty Charts. It seems probable that the description of 'Capo Sagres' in the next chapter really applies to the 'Capo de Verga', for the *Africa Pilot* (Pt. I, p. 252) says: 'Cape Verga cannot be mistaken, for, unlike all the adjacent coast, it rises at once from its base to a considerable elevation, and when seen from the southward, in connection with a conspicuous range of hills, is one of the most remarkable landmarks on the whole coast.... When Cape Verga bears 070°, distant thirteen or fourteen miles, two conical elevations, lying about one and a half miles northward of the Cape, will be seen; the eastern of these is isolated and shaped like a sugar loaf....'

[3] This is the approximate distance between the Rio Grande (Jeba) and Cape Verga.

[4] This is the promontory on the south-east of Sangaria Bay, off which lies Tumbo Island, on which Konakri stands. See the note on Cape Verga above. It may be, however, that the diamond-shaped point here referred to is the conical peak of Mount Kakulima.

was built by the Infante Don Hurich above one of the points of Capo San Vincenzo, called Sagres. For this reason it is known to the Portuguese as Capo di Sagres di Signea [Guinea]. The sailors say, from the information they gathered, that the inhabitants are idolators, and worship statues of wood in the form of men. They also say that when they are about to eat or drink, they first make offerings to their idols. They are more nearly brown in colour than black. They have marks made with hot irons on their faces and bodies. They always go naked, and in the place of loincloths wear the bark of trees to cover their private parts. They have no weapons, for iron is not to be found in their country. Their food is rice, millet, vegetables, such as beans, of different species to ours, for they are larger. They have the flesh of cows and goats, but not in large quantities.

Further, they say that off this cape there are two islets[1], the one six, the other eight miles off: on account of their small size they are uninhabited, but have plenty of very beautiful green trees.

Item the people of this river[2] have very big canoes (that is, what we style *zopoli*) which carry thirty to forty men. They ply their oars standing [without rowlocks], as I have said above.

These people all have their ears pierced round with holes in which they wear various gold rings, one behind the other. They also pierce their nostrils. When they wish to eat, they have to draw them aside. These are worn by men and women alike. They say also that the wives of the King and the chiefs, or men of standing, all have their lips pierced as well as their ears, in which to signify their grandeur and rank they wear rings of gold, and these they wear or lay aside as it pleases them.

About forty miles beyond Capo Sagres there is another river, called San Vincenzo[3]: its mouth is about forty [four] miles wide. Fifty [five] miles farther along the same coast, there is another

---

[1] The Isles do Los (or Idoles). Though, including Tumbo Island, they are seven in number, two are distinctly larger than the others. The preceding remarks appear to apply to their inhabitants.

[2] The river referred to here, though not mentioned elsewhere, is presumably either the Dembia or the Dubreka, which flow into Sangaria Bay.

[3] The Forikaria River.

river called Rio Verde¹, the mouth of which is wider than that of the Rio de San Vincenzo, that is, more than forty miles. These names were given to the rivers by the above-mentioned navigators in the King's caravels. All this region and coast is mountainous, with good harbours and anchorages.

About twenty-four miles beyond the Rio Verde there is another Cape to which they gave the name Capo Liedo², that is in our tongue 'Aliegro': because it seemed to them that this cape, with the surrounding country, was exceedingly pleasant.

Beyond this Capo Aliegro begins a mountainous coast which extends for about six [fifty] miles³, and is very high, covered with tall and perennially green trees: at the end of it, about eight miles out to sea, there are three islets, the largest of which is about ten to twelve miles in circumference. They named this the 'Isola Salvaza'⁴, and the mountain 'Montagna Liona'⁵ [Sierra Liona, because of the great noise of the thunder claps which are heard continually from the clouds which always surround the summit].

Beyond the coast of the 'Montagna Liona' all is low land, and fringed with many sandbanks which run out to sea⁶.

Thirty miles beyond the cape of this mountain there is another large river, three miles wide, which they named 'Fiume Rosso'⁷, because the river water appeared red, as the bottom was red

---

¹ Ramusio's emendation to 'five miles' makes this river the Mellakori.
² This distance is twenty-four miles in the manuscript; the printed version is again erroneous, probably due to the compositor mistaking 'xxiv' for 'cciv'. The emendation makes this 'Capo Liedo' Bullo Point; from the statement in the next paragraph and Duarte Pacheco's description, however, it is apparent that 'Capo Liedo' is the modern Cape Sierra Leone.
³ The extent of the Sierra Leone peninsula is about twenty miles.
⁴ Dublin Island, the largest of the three islands forming the Banana Islands.
⁵ Sierra Leone: Ramusio has substituted 'Sierra Liona' for 'Montagna Liona', and inserted the explanation. Duarte Pacheco, however, gives another version: 'Many believe that Serra Lyoa is so-called because there are lions here, but this is not so, for Pero de Sintra, a knight of Prince Henry's household, who discovered Serra Lyoa at the prince's bidding, seeing that it was a wild rough country, called it the Lioness, and there was no other reason; and there can be no doubt of this, for he told me so himself' (*Esmeraldo*, Hakluyt Soc. series II, LXXXIX, p. 99).
⁶ Yawri Bay and the Shoals of St Anne.
⁷ Cockboro River.

ground. Beyond this river there is a cape, the earth of which is red, to which they also gave the name Capo Rosso[1]. Off this cape, about eight miles out to sea, there is a red islet, originally part of this Capo Rosso, the 'Isola Rossa'[2]. At this island, the Pole Star appears at the height of a man above the sea. Note that from the mouth of the Fiume Rosso to this islet is about ten miles. Beyond the Capo Rosso there is a gulf[3], at the head of which is a large river. They named this 'Rio de Sancta Maria' because it was discovered by them on the day of Sancta Maria de la neue[4].

On the further side of the river there is a point[5] with an islet a short distance out to sea. In this gulf there are many sandy shallows which extend for twelve miles along the coast and over which the sea breaks. The sea currents are very strong, and there are great flood and ebb tides. To this islet they gave the name 'Isoletta di scanni'[6], on account of these banks.

Beyond the islet there is a great cape which they named 'Capo de Sancta Anna'[7], because I believe they discovered it on that day. From the island to the cape is about twenty-four miles: all this coast is shallow and has few anchorages.

[1] Shenge Point: 'is composed of soft red sandstone which is rapidly being encroached upon by the sea' (*Africa Pilot*, Pt. I, p. 276).
[2] One of the Plantain Islands.
[3] The Sherbro River, or, more properly, Sound, into which several rivers flow.
[4] The Bagru River.
[5] The modern Cape St Anne. Saint Anne's Day is July 26, and 'Our Lady of the Snow' August 5th; the text appears to be confused here.
[6] One of the Turtle Islands.
[7] If the distance from the Turtle Islands and the point beyond Bagru River is twenty-four miles as given, it would appear that the original 'Capo de Sancta Anna' was the promontory at the south-east of Sherbro Island, now called Argyle Point. The modern Cape St Anne is described as low and sandy: this does not seem to agree with the 'great cape' of the text.

CHAPTER L

The RIVER *of the* PALMS, *and many others*

Seventy miles along the coast beyond this Capo de Sancta Anna, there is another river, to which they gave the name 'Fiume de le Palme'[1], for there were many palms there. The mouth of this river, although it was seen to be of considerable width, is filled with shoals and sandbanks, so that the entry is very dangerous. From the Capo de Sancta Anna to this river it is shallow.

Having passed about seventy miles beyond this river, despite the shallows of the coast, they discovered a small river which they called the 'Rio de li Fiumi'[2], because at the time of the discovery they saw along the entire coast nothing but smoke over the land, made by the people of the country. Twenty-four miles beyond this river, still navigating the shallows, they found a cape which ran far out to sea: above it appeared a high mountain, wherefore they called it 'Capo del Monte'[3].

Beyond this Capo del Monte, having progressed about sixty miles along shore, they found a small, low cape, which also showed a hillock upon it: this they called 'Capo Cortese' [or 'Misurado'][4]. Here, on that first night, they saw many fires among the trees and along the shore, made by the Blacks on sighting the first Christian vessels which had ever been seen by them.

Sixteen miles along the shore beyond this cape, there is a great forest of very green trees which grow right down to the water's edge. [This they named the 'Bosco', or rather 'Arboreto', di Santa Maria.] To the caravels which anchored beyond this, there came some small canoes, two or three Blacks

---

[1] Sulima River: the description of the mouth of this river given in the text is very accurate. As it is approximately seventy miles from the south-east end of Sherbro Island, this is additional evidence that the original Cape St Anne has been transferred to its present position.

[2] This sentence appears to have been misplaced in the text. Twenty-four miles from the Sulima River carries one to the 'Capo del Monte', as is stated in the next sentence; seventy miles from the Sulima River is the St Paul River, close to Cape Mesurado.

[3] Cape Mount.   [4] Cape Mesurado.

in each, quite naked and carrying pointed sticks, what we should style darts. Some had small knives, and two leathern shields with three bows amongst them. They came alongside. Each had his ears pierced right round, and the lower part of the nose. Some of them also wore teeth around their necks; these appeared to be human teeth. Several negroes who were on the ships spoke to them, but without understanding a single word, nor making themselves understood. Three of the Blacks boarded one of the caravels: of these three, the Portuguese detained one, allowing the others to go. This they did in obedience to His Majesty the King, who had enjoined them that, from the farthest land they reached, if it chanced that the people were unable to understand their interpreters, they were to contrive to bring away a negro, by force or persuasion, so that he might be interrogated through the many negro interpreters to be found in Portugal, or in the course of time might learn to speak, so that he might give an account of his country. For this reason, they detained one of the three negroes. Deciding to advance no farther, they carried him back to Portugal and presented him to His Majesty the King, who caused divers negroes to speak with him. Finally a negress, the slave of a Lisbon citizen, who had also come from a far off country, understood him, not through his own language but through another known to both. What this negro told the king through this woman I do not know, save that he said that among other things found in his native country were live unicorns. The said lord, having kept him several months, shown him many things of his Kingdom, and given him some clothes, very charitably had him carried by a caravel back to his own country.

No ship had passed beyond this farthest place down to the time of my leaving Spain, which was on the first of February 1463. To the above mentioned wood was given the name 'Boscho de Sancta Maria'.

Finis

THE LETTER OF ANTOINE MALFANTE

Copy of a letter written from Touät by ANTOINE MALFANTE and addressed to GIOVANNI MARIONO at Genoa. 1447.

AFTER we had come from the sea, that is from Hono[1], we journeyed on horseback, always southwards, for about twelve days. For seven days we encountered no dwelling—nothing but sandy plains; we proceeded as though at sea, guided by the sun during the day, at night by the stars. At the end of the seventh day, we arrived at a *ksour*[2], where dwelt very poor people who supported themselves on water and a little sandy ground. They sow little, living upon the numerous date palms. At this *ksour* we had come into Tueto[3]. In this place there are eighteen quarters, enclosed within one wall, and ruled by an oligarchy. Each ruler of a quarter protects his followers, whether they be in the right or no. The quarters closely adjoin each other and are jealous of their privileges. Everyone arriving here places himself under the protection of one of these rulers, who will protect him to the death: thus merchants enjoy very great security, much greater, in my opinion, than in kingdoms such as Themmicenno[4] and Thunisie[5].

Though I am a Christian, no one ever addressed an insulting word to me. They said they had never seen a Christian before. It is true that on my first arrival they were scornful of me, because they all wished to see me, saying with wonder 'This Christian has a countenance like ours'—for they believed that

---

[1] Honein.
[2] Tabelbert.
[3] Tuat, a group of oases of which Tamentit is the capital.
[4] Tlemcen.
[5] Tunis.

Christians had disguised¹ faces. Their curiosity was soon satisfied, and now I can go alone anywhere, with no one to say an evil word to me.

There are many Jews, who lead a good life here, for they are under the protection of the several rulers, each of whom defends his own clients. Thus they enjoy very secure social standing. Trade is in their hands, and many of them are to be trusted with the greatest confidence.

This locality is a mart of the country of the Moors, to which merchants come to sell their goods: gold is carried hither, and bought by those who come up from the coast. This place is De Amamento², and there are many rich men here. The generality, however, are very poor, for they do not sow, nor do they harvest anything, save the dates upon which they subsist. They eat no meat but that of castrated camels, which are scarce and very dear.

It is true that the Arabs with whom I came from the coast brought with them corn and barley which they sell throughout the year at 'f. saracen, la nostra mina'³.

It never rains here: if it did, the houses, being built of salt in the place of reeds, would be destroyed⁴. It is scarcely ever cold here: in summer the heat is extreme, wherefore they are almost all blacks. The children of both sexes go naked up to the age of fifteen. These people observe the religion and law of Muhammad. In the vicinity there are 150 to 200 *ksour*.

In the lands of the blacks, as well as here, dwell the Philistines⁵, who live, like the Arabs, in tents. They are without number, and hold sway over the land of Gazola⁶ from the

---

¹ The Latin word employed is 'contrafactum'.   ² Tamentit.
³ A *Saracen* was the Arab coin known as the *dinar*; *Mina* was a measure equalling approximately half a bushel. The 'f.' perhaps stands for 'six'.
⁴ De la Roncière (1, p. 152) suggests that this is not a description of Tamentit but of Taghaza, where the houses were all built of rock salt.
⁵ The Tuareg.
⁶ Gazola (the *Gazula* of Idrisi) appears on many portolan charts, sometimes applied to a town, sometimes to a region. On the Pizigani chart of 1373 there is a cape, probably Cape Nun, called 'Caput finis Gozole'. It has been derived from the Berber people, the Guezulah, a branch of which inhabited the Sus; as used by Malfante it appears to be applied to the same area as 'Sarra', or Sahara, of other contemporary writers (see Magnaghi, A., *Precursori di Colombo?*, p. 45).

borders of Egypt to the shores of the Ocean, as far as Massa and Safi, and over all the neighbouring towns of the blacks. They are fair, strong in body and very handsome in appearance. They ride without stirrups, with simple spurs. They are governed by kings, whose heirs are the sons of their sisters—for such is their law. They keep their mouths and noses covered. I have seen many of them here, and have asked them through an interpreter why they cover their mouths and noses thus. They replied: "We have inherited this custom from our ancestors." They are sworn enemies of the Jews, who do not dare to pass hither. Their faith is that of the Blacks. Their sustenance is milk and flesh, no corn or barley, but much rice. Their sheep, cattle, and camels are without number. One breed of camel, white as snow, can cover in one day a distance which would take a horseman four days to travel. Great warriors, these people are continually at war amongst themselves.

The states which are under their rule border upon the land of the Blacks. I shall speak of those known to men here, and which have inhabitants of the faith of Muhammad. In all, the great majority are Blacks, but there are a small number of whites [i.e. tawny Moors].

First, Thegida[1], which comprises one province and three *ksour*; Checoli[2], which is as large.

Chuchiam[3], Thambet[4], Geni[5], and Meli[6], said to have nine towns;

Thora[7], Oden[8], Dendi[9], Sagoto[10], Bofon[11], Igdem[12], Bembo[13], all these are great cities, capitals of extensive lands and towns under their rule.

These adhere to the law of Muhammad.

To the south of these are innumerable great cities and territories, the inhabitants of which are all blacks and idolators, continually at war with each other in defence of their law and

---

[1] Takedda, five days' march west-south-west of Agadez.
[2] Possibly Es Suk (Tadmekka), north of Takedda at the head of the Tilemsi valley.
[3] Probably Gao.   [4] Timbuktu.   [5] Jenne.
[6] Mali.   [7] Unidentified.   [8] Wadan.
[9] Dendi, probably the original home of the Songhai.
[10] Unidentified.   [11] Unidentified.   [12] Unidentified.
[13] Possibly Bamba, a town on the Middle Niger.

faith of their idols. Some worship the sun, others the moon, the seven planets, fire, or water; others a mirror which reflects their faces, which they take to be the images of gods; others groves of trees, the seats of a spirit to whom they make sacrifice[1]; others again, statues of wood and stone, with which, they say, they commune by incantations. They relate here extraordinary things of this people.

The lord in whose protection I am, here, who is the greatest in this land, having a fortune of more than 100,000 *doubles*, brother of the most important merchant in Thambet, and a man worthy of credence, relates that he lived for thirty years in that town, and, as he says, for fourteen years in the land of the Blacks. Every day he tells me wonderful things of these peoples. He says that these lands and peoples extend endlessly to the south: they all go naked, save for a small loin-cloth to cover their privates. They have an abundance of flesh, milk, and rice, but no corn or barley.

Through these lands flows a very large river[2] which at certain times of the year inundates all these lands. This river passes by the gates of Tambet, and flows through Egypt; and is that which passes by Carium[3]. There are many boats on it, by which they carry on trade. It would be possible, they say, to descend to Egypt by this river, were it not that at a certain spot it falls 300 cubits over a rock[4], on account of which boats cannot go or return. This river flows at about twenty days' journey on horseback from here.

These people have trees which produce an edible butter[5], of which there is an abundance here. I have seen them bearing it hither: it is as wonderful an unguent as the butter of sheep. The

---

[1] Cf. Seligmann (*op. cit.* pp. 61–2). 'The Bambara have been little affected by Islam and retain their animistic beliefs and ancestor worship. Each village has its presiding spirit (*dasiri*) or divine ancestor, usually resident in a tree at which sacrifices are made and prayers offered by the *dugutigi* on all important occasions.'

[2] The Niger. It is to be noted that Malfante correctly implies that it flows eastwards, of which there was no certain knowledge till the end of the eighteenth century.

[3] Cairo. On the confusion between the Niger and the Nile, see p. 28 above.

[4] This appears reminiscent of the cataracts of the Nile.

[5] The Karité tree, the most characteristic of the Savannah. The butter is obtained from the kernel.

slaves which the blacks take in their internecine wars are sold at a very low price, the maximum being two *doubles* a head. These peoples, who cover the land in multitudes, are in carnal acts like the beasts; the father has knowledge of his daughter, the son of his sister. They breed greatly, for a woman bears up to five at a birth. Nor can it be doubted that they are eaters of human flesh, for many people have gone hence into their country. Neither there nor here are there ever epidemics.

When the blacks catch sight of a white man from a distance, they take to flight as though from a monster, believing him to be a phantom. They are unlettered, and without books. They are great magicians, evoking by incense diabolical spirits, with whom, they say, they perform marvels.

"It is not long since I was in Cuchia[1], distant fifty days' journey from here, where there are Moors", my patron said to me. "A heathen king, with five hundred thousand men, came from the south to lay siege to the city of Vallo. Upon the hill within the city were fifty Moors, almost all blacks. They saw that they were by day surrounded by a human river, by night by a girdle of flames and looked upon themselves as already defeated and enslaved. But their king, who was in the city, was a great magician and necromancer; he concluded with the besieger a pact by which each was to produce by incantation a black goat. The two goats would engage in battle, and the master of that which was beaten, was likewise to consider himself defeated. The besieger emerged victorious from the contest, and, taking the town, did not allow one soul to escape, but put the entire population to the sword. He found much treasure there. The town to-day is almost completely deserted save for a poverty-stricken few who have come to dwell there."

Of such were the stories which I heard daily in plenty. The wares for which there is a demand here are many: but the principal articles are copper, and salt in slabs, bars, and cakes. The copper of Romania[2], which is obtained through Alexandria, is always in great demand throughout the land of the Blacks. I frequently enquired what they did with it, but no one could

[1] Gao.
[2] The Byzantine Empire.

give me a definite answer. I believe it is that there are so many peoples that there is almost nothing but is of use to them.

The Egyptian merchants come to trade in the land of the Blacks with half a million head of cattle and camels—a figure which is not fantastic in this region.

The place where I am is good for trade, as the Egyptians and other merchants come hither from the land of the Blacks bringing gold, which they exchange for copper and other goods. Thus everything sells well; until there is nothing left for sale. The people here will neither sell nor buy unless at a profit of one hundred per cent. For this reason, I have lost, Laus Deo!, on the goods I brought here, two thousand *doubles*.

From what I can understand, these people neighbour on India[1]. Indian merchants come hither, and converse through interpreters. These Indians are Christians, adorers of the cross. It is said that in the land of the Blacks there are forty dialects, so that they are unable to understand each other.

I often enquired where the gold was found and collected; my patron always replied "I was fourteen years in the land of the Blacks, and I have never heard nor seen anyone who could reply from definite knowledge. That is my experience, as to how it is found and collected. What appears plain is that it comes from a distant land, and, as I believe, from a definite zone." He also said that he had been in places where silver was as valuable as gold.

This land is twenty-eight days' journey from Cambacies[2], and is the city with the best market. It is twenty-five days from Tunis, from Tripoli in Barbary twenty days, from Trimicen[3] thirty days, from Fecia[4] twenty days, from Zaffi[5], Zamor[6] and Messa twenty days on horseback. I finish for the present; elsewhere and at another time, God willing, I will recount much more to you orally. I am always at your orders in Christ.

Your ANTONIUS MALFANT

[1] Probably Abyssinia, the kingdom of Prester John, which was regarded in the Middle Ages as one of the Three Indias.
[2] Probably Ghadames.   [3] Tlemcen.   [4] Fez.
[5] Safi.   [6] Azamor.

THE
VOYAGES *of* DIOGO GOMES

SOME time after this[1] the Prince equipped at Lagos a caravel, named *Picanso*, and appointed Diogo Gomes captain, together with two other caravels, of which he appointed Diogo Gomes captain-in-chief. The captain of one of these was João Gonçalves Ribeiro, of the Prince's household, and of the other Nuño Fernandes de Baya, the Prince's esquire-at-arms. The Prince gave them orders to proceed as far as they could. After passing the river of S. Domingos[2], and another great river called Fancaso[3] beyond the Rio Grande, we encountered strong currents in the sea, so that no anchor could hold. These are called *macareo*[4]. The other captains, therefore, and their men were greatly alarmed, thinking that they were at the extremity of the ocean, and they begged me to return. In the middle of the current the sea was very clear, and the Mouros came from the shore in their canoes and brought us their merchandise, viz., cotton cloth, or *algodão*, elephants' teeth, and a quart measure of *malagueta*, in grain and in its pods as it grows, with which I was much delighted. The current prevented our proceeding farther, and in fact increased so much that it obliged us to put back.

We came to a land where, near the shore, were many palm trees, with their branches broken, and so tall that from a

[1] The unsuccessful expedition of Adalbert, or Vallarte, the Dane, on which he lost his life (see Azurara, *op. cit.* II, p. 280). This took place in 1448; from internal evidence, Gomes's voyage took place in 1457, or perhaps 1456 (see Introduction, p. xxiv).
[2] The Mansoa River: the Rio de San Dominico of Cadamosto.
[3] The present Rio Grande. On Benincasa's map of 1468, 'Fancaso' appears to be used for Cadamosto's 'Rio Grande' (the Jeba). Jobson says 'Fancassa' was the native name for the sea: '...the Sea, whereof they are altogether ignorant, onely by the name, or word Fancassa, which signifieth great waters' (*Golden Trade*, p. 93).
[4] Cadamosto (p. 76) also refers to these currents.

distance we thought that they were masts or spars of negro vessels[1]. Thither we went, and found an extensive plain full of grass, and more than five thousand animals called in the negro language *myongas*. These are beasts a little larger than stags, which showed no fear at sight of us. We also observed five elephants come out of a small river sheltered with trees. Three of them were large, with two young ones, and they fled from the *myongas*. On the sea-shore we saw many crocodiles' holes. We returned to the ships, and on the next day made our way from Cape Verde, and we saw the broad mouth of a river, three leagues in width, which we entered, and from its size correctly concluded that it was the river Gambia. We entered it with the wind and tide in our favour, and came to a small island in the middle of the river[2], and there remained that night. In the morning, however, we went farther in, and saw many canoes full of men, who fled at sight of us, for it seems they were the same who had slain Nuño Tristão and his men[3]. The next day, however, we saw beyond the head (? point) of the river some people on the right hand bank, to whom we went, and were received in a friendly manner. Their chief was called Frangazick, and was the nephew of Farisangul, the great Prince of the negroes. There I received from the negroes one hundred and eighty *arrateis* weight of gold, in exchange for our merchandise; such as cloths, necklaces, &c. They told us that the negroes on the left shore would not hold intercourse with us because they had slain the Christians. The lord of that country had a certain negro, named Bucker, who was acquainted with the whole country of the negroes, and finding him perfectly truthful, I asked him to go with me to Cantor, and promised to give him a mantle and shirts, and every necessary. I made also a similar promise to his chief, which I kept. We ascended the river, and I sent a captain with his caravel into a certain harbour, named

[1] This appears to be the 'Cabo de Mastos', identified by A. Cortesão with Red Cape, just beyond Cape Verde. If so, Gomes had 'put back' a considerable distance, but the later reference to Cape Verde confirms this identification.

[2] This may be Cadamosto's 'Isola de Santa Anna' (see p. 66).

[3] Nuno Tristão was killed on the Rio Grande. It has been suggested that Gomes is confusing the attack on Cadamosto on his first voyage with this earlier event.

Ulimays. The other remained in Animays[1], and I went up the river as far as Cantor[2], which is a large town near that river's side. On account of the thick growth of the trees on both sides of the river, the vessels could proceed no farther, and I sent out the negro whom we had brought with us, to make it known to the people of the country that I had come thither for the purpose of exchanging merchandise, and, in consequence, the negroes came in very great numbers. When the report spread throughout the country round, that the Christians were in Cantor, the natives came together from all quarters, viz., from Tambucutu in the north, from the Serra Geley in the south, and there came also people from Quioquun[3], which is a great city, surrounded by a wall of baked tiles, and where I understood there was abundance of gold, and that caravans of camels and dromedaries crossed over thither with merchandise from Carthage or Tunis, from Fez, from Cairo, and from all the land of the Saracens, in exchange for gold. They said that the gold was brought from the mines of Mount Gelu, which is the opposite side of the range called Sierra Lyoa[4]. They said that that range of mountains began at Albafur[5], and ran southwards, which pleased me much, because all the rivers, large and small, descending from those

[1] Major suggests that this is a corruption of Gnumimansa.

[2] Gomes says later that Cantor was some fifty leagues up the Gambia. It will be remembered that Cadamosto's parley with Battimansa took place about sixty miles up stream. Cantor is mentioned by Duarte Pacheco as a district 150 leagues upstream. Jobson says: 'On the southside [of the Gambia] the whole country we past, even to the highest we went, which you must needes conceive to be very spacious, had all reference to the great King of *Cantore*' (*Golden Trade*, p. 60). Barros places the trading station 180 leagues up the Gambia.

[3] Gao on the Niger (see Introduction, p. xvii). Gomes should be taken as implying that Cantor was the terminus of trade routes from the towns he mentions, and not that traders came thither on learning of his arrival.

[4] This appears to be the first recognition of the existence of a divide between the westward and eastward flowing rivers of western Africa. The mountains referred to are the Futa Jallon, in which the upper streams of the Niger rise. Sierra Leone was not of course discovered by de Sintra for some years after the probable date of Gomes's voyage. This information was no doubt acquired by Gomes later, but incorporated in his narrative.

[5] The identity of this range of mountains 'which began at Albafur' is open to doubt. It would appear from the preceding statement that it formed the watershed of the rivers flowing to the Atlantic, and this is the interpretation accepted by Prestage (*Portuguese Pioneers*, p. 131). The difficulty is raised by the use of the name Albafur in this connection. Both Duarte

mountains (which had been as yet observed) ran westward; but they told me that other very large rivers ran eastward from them, and that near that city was a certain great river, named Emin[1], and that there was also a great lake, but not very broad, on which were many canoes, like ships, and that the people on the opposite sides were in constant warfare with each other, those on the eastern side being white men[2]. On my enquiring what sovereigns ruled in those parts, they answered that the chief of that part, which was inhabited by the negroes, was named Sambegeny, and that the lord of the eastern part was called Semanagu, and that a short time before they had a great battle, in which Semanagu was the conqueror. And a certain Saracen of Termezen[3], named Admedi, told me that he had been through all that land, and had been present at the battle,

Pacheco and Rodriguez refer to mountains of Baffor, or Bafoor, much farther to the north. Kimble, in his edition of Duarte Pacheco (p. 76), speaks of the Bafur as 'a sedentary folk inhabiting the Sudan fringe of the Sahara', though he concludes with Prestage that the mountains are to be located in the hinterland of Sierra Leone. But the evidence of Rodriguez, a factor at Arguim, is even stronger in favour of a northerly location. In his description of the province of 'Lodea', the coast land of Arguim, divided from Guinea by the Senegal, the following passage occurs: 'Six miles from the rock Schelud lies a very high mountain, called Baffor. It is eighteen miles long, and as steep as a wall, especially on the north towards the desert. This mountain is so lofty that it seems to reach the sky. It has three approaches only, which appear unclimbable, but which are accessible to camels. In this mountain lie the approaches to four cities, one market town, and fourteen waterless streams. It is quite overgrown with tamarinds. Of these cities, the most important is called Oadem, the second Oulili, the third Schinquete, the fourth Tunniqui; the market town is called Fara' (Kunstmann, F., 'V. Ferdinands Beschreibung', *Abh. hist. Cl. k. bay. Ak. Wissens.*, VIII (1856), i, p. 270). It is clear, therefore, that the mountains of Albafur lay to the north of the Senegal, and 'Schinquete' is evidently Chinguetti in Adrar. In assuming that the range at the head of the Gambia was related to the northern, Gomes is supported by modern observations: cf. *Atlas des colonies françaises*: 'En retrait se dressent en falaises les grès primaires. Ils atteignent une altitude voisine de 500 mètres dans l'Adrar mauritanien (Atar, Chinguetti). Les assises de l'Adrar se prolongent au nord par le Hank. Au sud, par le Tagant, etc., on les suit jusqu'au Fouta-Djallon' (p. 47).

[1] The 'certain great river' must be the Niger, though I have not encountered the name Emin elsewhere: possibly it is connected with the Amenuam or Emenuam of the Spanish Friar. Duarte Pacheco, however, notes that with the Bulom of Sierra Leone the word for water was 'men' (*op. cit.* p. 97). The 'great lake' is the inundated area of the Niger, which figures in Cadamosto's and other accounts of Wangara.

[2] I.e. tawny Moors, in contradistinction to blackamoors.

[3] Tlemcen.

both by sea and land. When I afterwards related all these things to the Prince, he told me that a merchant in Oran had written to him two months before respecting this engagement, which had taken place between Semanagu and Sambegeny, and, therefore, he believed the account. Such are the things which were told me by the negroes who were with me at Cantor. I questioned the negroes at Cantor as to the road which led to the countries where there was gold, and asked who were the lords of that country. They told me that the king's name was Bormelli, and that the whole land of the negroes on the right side of the river was under his dominion, and that he lived in the city Quioquia. They said further, that he was lord of all the mines, and that he had before the door of his palace a mass of gold just as it was taken from the earth, so large that twenty men could scarcely move it, and that the king always fastened his horse to it, and kept it as a curiosity on account of its being found just as it was, and of its great size and purity[1]. The nobles of his court wore in their nostrils and ears ornaments of gold. They said also, that the parts to the east[2] were full of gold mines, and that the men who went into the pits to get the gold did not live long, on account of the impure air. The gold-bearing sand was afterwards given to women to wash the gold from it.

I inquired the road from Cantor to Quioquia, and was told that to Morbomelli[3] from Cantor the road is eastward to Somanda, and from Somanda to Commuberta and to Cereculle[4] and other places, the names of which I have forgotten. And in these aforenamed places is great abundance of gold, as I can well believe, for I saw the negroes at the time who went by these roads come loaded with gold. And they said that Forisangul[5] was subject to Mormelli, who was lord of the right bank of the river Gambia.

[1] In the twelfth century Idrisi (ed. Dozy and de Goeje, p. 8) reported that the royal treasure of the then powerful Soninke kingdom of Ghana, lying west of Timbuktu, included a nugget of gold to which the king was accustomed to tie his horse.
[2] The Wangara area, east of the Gambia.
[3] Kukia or Gao.
[4] According to Barros, the Caragoles were a people living on the upper Senegal.
[5] This is in accord with Cadamosto's statement, see p. 67.

While thus holding peaceful intercourse with these negroes of Cantor, my men became worn out with the heat, and so we returned in search of the other two caravels, and in the caravel which had remained in Ollimansa, I found nine men had died, the captain, Gonçalo Affonso, very ill, and all the rest of his men sick, except three. I found the other caravel fifty leagues lower down towards the ocean, and in it five men had died. We immediately withdrew, and made for the sea, and I went to the place where I had hired the negro traveller, and gave him what I had promised him.

They then informed me that on the other, that is, the left or south side of the river, there was a certain great chief named Batimansa, and I desired to make peace with him, and I sent to him that negro who had been with me at Cantor. That chieftain desired to speak with me in a great wood on the bank of the river, and brought with him an immense throng of people armed with poisoned arrows, azagays, and swords and shields (*adagas*). And I went to him, carrying him some presents and biscuit, and some of our wine, for they had no wine except what is made from the date palm[1], and he gave me three negroes, one male and two female, and he was pleased and extremely gracious, making merry with me and swearing to me by the one true God that he would never again war against the Christians[2], but that they might travel safely through his land and exchange their merchandise. Being desirous of putting this to the proof, I sent a certain Indian named Jacob, whom the Prince had sent with us, in order that, in the event of our reaching India[3], he might be able to hold speech with the natives, and I ordered him to go to the place which is called Alcuzet, with the lord of that country, whither, on a former occasion, a knight had gone with him, through the land of Geloffa to find the land of Gela and Tambucutu. This Jacob, the Indian, related to me that Alcuzet[4]

---

[1] Gomes is referring to palm wine, called by Cadamosto *Miguol*.
[2] Cadamosto had been amicably received by Battimansa on his second voyage.
[3] See p. 90, note.
[4] I have not identified this place with certainty. Leo mentions 'Alguechet', an oasis on the eastern side of the Sahara, but though it is stated that its inhabitants traded with the land of the Blacks, it can scarcely be the place in question.

is a very vicious land, having a river of sweet water and abundance of lemons, some of which he brought with him to me. And the lord of that country[1] sent me elephants' teeth, one of them very large, and four negroes, who carried the tooth to the ship. And so they came peacefully up to our ships, and thus I was safe from them. Afterwards I went to his abode, which was surrounded by many negro habitations. Their houses are made of seaweeds, covered with straw, and I remained with them for three days. Here were many parrots and many ounces, and he gave me six skins, and ordered that an elephant should be killed and its flesh carried on board the caravels.

It was here that I learned the fact that all the mischief that had been done to the Christians had been done by a certain king, called Nomymans[2], who possesses the land which lies on that promontory. I took great pains to make peace with him, and sent him many presents by his own men in his own canoes, which were going for salt to his own country. This salt was plentiful there, and of a red colour. He greatly feared the Christians, on account of the injury which he had done them. I went by the river towards the ocean, as far as the harbour, near the mouth of the river, and he sent to me many times men and women to try me, whether I would do them any harm, but, on the contrary, I always gave them a friendly reception. When the King heard this, he came to the river-side with a great force, and sitting down on the bank, sent for me to come to him, which I did, paying him all ceremonious respect in the best fashion I could. There was a certain Bishop there of his native church, who put questions to me with respect to the God of the Christians, and I answered him according to the intelligence which God had given me, and at last I questioned him respecting Muhammad, in whom they believe. What I said pleased the King so much, that he ordered the Bishop within three days to take his departure out of his kingdom, and springing to his feet, he declared that no one, on pain of death, should dare any more to utter the name of Muhammad, for he believed in the one God only, and that there was no other God but He, in

---

[1] Battimansa, not the lord of Alcuzet.
[2] ? Gnumimansa, see page 71.

whom his brother, the Prince Henry, said that he believed. Calling the Infante his brother, he desired that I should baptize him, and so said also all the lords of his household, and his women likewise. The King himself declared that he would have no other name than Henry, but his nobles took our names, such as Jacob, Nuño, &c., as Christian names. I remained that night on shore with the King and his chiefs, but I did not dare to baptize them, because I was a layman.

On the next day, however, I begged the King with his twelve principal chiefs and eight of his wives to come to dine with me on board the caravel, which they all did unarmed, and I gave them fowls and meat cooked after our own fashion, and wine, both white and red, as much as they pleased to drink; and they said to each other that no nation was better than the Christians.

Afterwards, when we were on shore, he desired that I would baptize him; but I answered that I had not received authority from the supreme pontiff. I told him, however, that if he so desired, I would convey his wishes to the Prince, who would send a priest to baptize them. He immediately wrote to the Prince to send him a priest, and someone to inform him respecting the faith, and begged the Prince to send him a falcon for hunting, for he wondered greatly when I told him that the Christians carried a bird on the hand which caught other birds. He wished him also to send two rams, and sheep, and ganders and geese, and a pig, as well as two men who would know how to construct houses and make a survey [?] of his city. All these requirements I promised that the Prince would fulfil. At my departure he and all his people lamented, so great was the friendship which had sprung up between him and me.

It so happened that for two years no one went back to Guinea, because King Affonso was gone, with a fleet of three hundred and fifty-two ships, to Africa, and took the powerful city of Alcacer dalquivi, for which reason the Prince, being fully occupied, gave no attention to Guinea. I then reminded the Lord Infante of the matters about which this king had written, and he ordered everything to be done as I had promised.

After leaving the king of Gambia I pursued my way to Portugal, and sent one caravel with those who were in the best health straight home. The other remained with me, because the people on board of her were sick. And I ordered the captain of the first vessel, if he had a favourable wind, to go straight to Portugal, if not, to wait for me at Arguim, and so he departed; but I with the other caravel sailed with a favourable wind to Cape Verde. As we came near the sea-shore we saw two canoes putting out to sea. We sailed between them and the land, and came up to them, and in one of the canoes we counted thirty-eight men, and the interpreter came to me, and said in my ear, that that was Beseghichi[1], lord of that land, a malicious man, of whom I have already spoken. I made them come into the caravel, and gave them to eat and drink, and a double portion of presents, and pretending that I did not know that he was the chief, said to him by way of trying him, "Is this the land of Beseghichi?" He said "Yes." I replied, "Why is he then so malignant against the Christians? It would be better for him to make peace with them, and that both might exchange merchandise, so that he might have horses, &c., as Burbuck and Badamel, and other lords of the negroes had. Tell him that I have taken you in this sea, and for love of him have set you free to go on shore." He much rejoiced, and I told them to go into their canoes, which they did, and as they all stood in their canoes, I said to the chief, "Beseghichi, Beseghichi, do not think that I did not know thee. It was certainly in my power to do with thee whatever I wished, and since I have acted kindly by thee, do thou do likewise with our Christians," and so they went their way.

A few days after we came to Cape Tofia, and to Anterot[2], and entered Arguim. And not far from the coast we came to the island called the Ilha de Garças. It was not inhabited, and was only one league in circumference. On it we found an innumerable multitude of birds of every kind, and on the ground pelicans' nests, and many dead pelicans. These are not as the painters represent them, but have a broad beak, and a

---

[1] The 'Isola di Beseghichi' is Gorée Island, near Cape Verde. According to Barros, Pero d'Evora made peace with Beseghichi on the voyage for the foundation of El Mina in 1481.

[2] The coast from Arguim to the Senegal.

stomach large enough to hold a measure of wheat, such as is commonly called an *alqueiro*. The number of birds there was so great, that we killed as many as we could carry in our boat, and so we went into Arguim. We thus sailed for Portugal, and came to Algarve, to the great city named Lagos, where the Prince at that time was, and he rejoiced greatly at our arrival.

After the Prince returned from the fleet with King Affonso, I reminded him of what King Nomimans had said, so that he should send to him all those things which had been promised. This the Prince did, and sent thither a certain priest, a relation of the cardinal's, the Abbot of Soto de Cassa, that he should remain with that king and instruct him in the faith. He also sent with him a young man of his household, named João Delgado. This was in the year 1458[1].

Two years afterwards[2] the Lord King Affonso equipped a large caravel, in which he sent me out as captain, and I took with me ten horses and went to the land of Barbacins[3], which is between Serreos[4] and King Nomimans. Those Barbacins had two kings, viz. Barbacin Dun and Barbacin Negor. And the King gave me authority over the shores of that sea, that whatever caravels I might find off the land of Guinea should be under my command and rule, for he knew that there were caravels there which carried arms and swords to the Moors, and he ordered me to take such prisoners and bring them to him to Portugal. And by the help of God in twelve days I arrived at Barbacins, and found there two caravels: one, in which was Gonçalo Ferreira, of the household of Prince Henry, a native of Oporto, who was conveying horses thither; and in the other caravel was a captain, Antonio da Noli[5], a Genoese, who was also a merchant conveying horses. This was in the port of Zaya[6]. I found there also Borgebil, who had been King of Geloffa, and who had fled thither from fear of the King Burbuck, who had

---

[1] The following portion, here omitted, deals with the death of Prince Henry.
[2] For the date of Gomes's second voyage, see Introduction, page xxvi.
[3] More accurately 'the land of the king of Sine'.
[4] The land of the Sereri.
[5] For da Noli, see Introduction, p. xxxix.
[6] Zaya may have been at the mouth of the Rio de Barbacini (the Joal).

taken his country from him. The aforesaid merchants with their caravels greatly damaged the traffic in those parts, for whereas the Moors used to give seven negroes for one horse, they gave them now no more than six. Then I summoned those captains to me, and on behalf of the King gave them seven negroes for one horse, but myself exchanged every horse for fourteen or fifteen negroes. While we were there, there came a caravel from Gambia, which brought us information that a certain man, named De Prado, was coming with a very richly laden caravel, whereupon I immediately fitted out the caravel of Gonçalo Ferreira, and ordered him in the King's name, on pain of death, and confiscation of all his goods, to go to Cape Verde, and to look out for that caravel, which he did, and took it, and we found great booty in it. I forthwith despatched the captain, together with Gonçalo Ferreira, to the King, and wrote to the King an account of all these events.

I and Antonio da Noli then left that port of Zaya, and sailed two days and one night towards Portugal, and we saw some islands in the sea[1], and, as my caravel was a lighter sailer than the other, I came first to one of those islands, and saw white sand, and it seemed to me a good harbour, and I cast anchor there, and so also did Antonio. I told them that I wished to be the first to land, and so I did. We saw no sign of any man there, and we called the island Santiago: it is so called to this day. There was abundance of fish to be caught there. On the shore we found many strange birds and streamlets of fresh water. The birds were so tame, that we killed them with sticks; and there were many geese there. There were also an abundance of figs, but they do not grow on the trees in the same manner as in our parts, for our figs grow near the leaf, but these all over the bark from the foot of the tree to the top. These trees grow in great numbers, and there was great quantity of grass there. And I had a quadrant when I went to these parts, and I wrote on the table of the quadrant the altitude of the Arctic Pole, and I found it better than the chart. It is true that the course of sailing is seen on the chart, but when once you get wrong, you

---

[1] On the question of the discovery of the Cape Verde Islands, see Introduction, p. xxxvi.

do not recover your true position[1]. We afterwards saw one of the Canary Islands, called Palma, and after that we went to the island of Madeira. Though I was anxious to go to Portugal, I was driven by a contrary wind to the Azores, but Antonio da Noli remained at Madeira, and availing himself of a more favourable wind, reached Portugal before me. And he begged of the King the captaincy of the Island of Santiago, which I had discovered, and the King gave it to him, and he kept it till his death. And I with extreme labour made my way to Lisbon, and after some time the King went to Oporto, where that De Prado, who had carried arms to the Moors, and whom Gonçalo Ferreira had taken prisoner, lay in irons, and the King ordered that they should martyrize him in a cart, and that they should make a furnace of fire, and throw him into it with his sword and gold.

[1] This is the earliest reference to the use of the quadrant for navigation. The passage appears to mean that Gomes observed the altitude of the Pole Star with his quadrant, then made the necessary correction, in accordance with the 'regiment of the North Pole' engraved upon the quadrant. The earliest chart now extant which bears a scale of latitudes is one by Pedro Reinel, *circa* 1505. It is possible that in Gomes' time some kind of sailing directions containing a list of prominent coastal features with their latitudes, similar to that given later by Duarte Pacheco, had already been compiled. (See Fontoura da Costa, p. 31.)

THE
ASIA *of* JOÃO DE BARROS[1]

THE FIRST DECADE
BOOK TWO

Of the deeds which the Portuguese performed in the discovery and conquest of the seas and lands of the Orient: containing what was achieved in the time of *King Afonso the Fifth* of that name in *Portugal*.

CHAPTER I

How the KING, D. AFONSO, *the* FIFTH *of that name,* SUCCEEDED *to the* GOVERNMENT *of this* KINGDOM, *at the end of his minority; although* PRINCE HENRY *pursued this discovery as long as he lived, we continue this* HISTORY *with the* KING, *and not with him: and of the* REASONS WHY *we* DO NOT WRITE *about other* FEATS *of this* KING.

When King Afonso, his minority ended, began to govern at the age of seventeen, he immediately sent ships to this discovery[2]. The Prince, however, on his part, also continued it, and at Santarem, on the second of September, 1448, the King[3] issued a charter in his favour, prohibiting others from promoting dis-

---

[1] João de Barros (1496–1570) was commander of S. Jorge da Mina from 1522 to 1525; Treasurer of the *Casa da India e Mina*, 1525–8, and Factor in 1532. The first decade of his *Asia* was published in 1552, the second in the following year, the third in 1563, and the fourth, a considerable time after his death, in 1613. Other works which he planned, or partially carried out, are mentioned below.

[2] Afonso V succeeded to the throne in 1438 as a minor; he assumed the government of the kingdom in 1449, and reigned until 1481. There is no record of these early voyages under his direction.

[3] These rights had already been conferred upon Prince Henry by the charter of October 22nd, 1443. He was also granted the monopoly of trade between Capes Cantim and Bojador by the charter of February 25th, 1449 (*Alguns documentos*, pp. 8, 13).

covery beyond Cape Bojador; and granting to him, as long as he wished, the fifth and the tithe of everything brought thence. This grant the Prince enjoyed until his death. But as at the beginning of the King's reign, there was strife between him and Prince Peter, the former Regent of these Kingdoms, which we related in the part of *Europa*[1], and also expeditions to Africa and Castille which occupied most of the King's reign, the course of this enterprise was not so continuous as it had been in the time of Prince Henry[2]. Gomes Eanes de Zurara[3], Chronicler of these Kingdoms, a man very skilled in the art of history, on which account he well deserved the charge he had, took care to record these feats; so that if there be anything well written in the Chronicles of this Kingdom, either of events that happened in his time or of those that happened before which had not been recorded, it came from his hand. What he wrote about these discoveries in the time of Prince Henry, he had received (according to what he says) from one Afonso Cerveira[4], the first to set these matters in order. We found some letters written at Beny by this Afonso Cerveira, when he was there as a factor of King Afonso. Though all, or the most part of what we have written so far, was taken from the writings of Gomes Eanes, and of Afonso Cerveira, we had no little work in putting together scattered materials and torn and disordered papers used by Gomes Eanes in the course of this discovery. Although he had promised to relate the events of the reign of King Afonso, we found no account of them, so that it seems as though he had the intention but not the opportunity; or if he wrote them, they are lost, as are so

---

[1] There are frequent references in the *Decades* to a work which Barros had planned but apparently never carried out. It was to have been a history of the Portuguese people, divided into three parts—*Europa*, *Africa*, and *Santa Cruz*—the first dealing with the early history of Portugal.

[2] This refers to the period after Prince Henry's death in 1460.

[3] Gomes Eanes de Zurara, or Azurara (*c*. 1400–*c*. 1473), the author of *The Chronicle of the discovery and conquest of Guinea*, which brings the story of Portuguese expansion down to the year 1448, the point at which this extract from Barros begins. The *Chronicle* was completed in 1453, and he was appointed Guarda Mór of the Torre do Tombo, and probably also Chronista Mor, in the following year. (*See* Beazley, C. R. and Prestage, E., *Chronicle of discovery of Guinea*, vol. I, Hakluyt Soc. xcv, pp. xiii ff.)

[4] Afonso Cerveira: author of a history of the Portuguese conquests on the coasts of Africa, now lost, which formed the basis of Azurara's chronicle.

SETTLEMENT OF THE AZORES

many other writings, which time has destroyed[1]. What we record therefore of the reign of King Afonso are no more than some memoirs which we found in the Tombo[2], and in the books of his exchequer,—not consecutive annals, as before, but fragments only of this discovery. From these memoirs we learn that in the year fourteen hundred and forty nine the King gave permission to the Prince to people the seven islands of the Azores, which had already been discovered[3], and on which, by order of the same Prince, one Gonçalo Velho, commander of Almourol, near the Town of Tancos, had landed cattle. And in the year fourteen hundred and fifty seven the King made a gift of all the islands then discovered[4] to his brother Prince Fernando, with civil and criminal jurisdiction, though with some limitations. And in fourteen hundred and sixty Prince Henry gave the two Islands of Jesus and Graciosa[5] to Prince Fernando, his nephew and adopted son, reserving for himself only the spirituality which belonged to the Order of Christ, of which he was governor. This gift was confirmed by the King in Lisbon on the second of September of the same year[6]. In the following year, as traffic in gold and negroes from Guiné was active in the Islands of Arguim, the King ordered the building of the Castle of Arguim[7], which stands there to-day, by Soeiro Mendes, a nobleman of his household, and inhabitant of Evora, to whom he gave the

[1] Barros later corrects this statement; see below p. 113, note.

[2] The national archives of Portugal are known as the *Archivo Nacional da Torre do Tombo* from a tower of that name in the Castello de São Jorge where they were kept from 1375. After the great earthquake of 1755 they were removed to their present home in the building now known as the Palacio das Cortes.

[3] Some of the islands of the Azores appear on maps of the fourteenth century; they were probably rediscovered by the Portuguese in 1427 and 1431–2. (See Arruda, M. M. V., *Descobrimento e povoamento dos Açôres*, Ponta Delgada, 1932.) The date of the license to people the islands should be 1439 (see the charter in *Alguns documentos*, p. 6).

[4] The donation of November 17th, 1457, conferred upon the Infant, D. Fernando, any islands discovered by his vessels *after* that date (*Alguns documentos*, p. 22).

[5] Islands of the Azores. Jesus Christo is now known as Terceira.

[6] The donation of December 3rd (not September 2nd), 1460 included also the islands of Madeira, Porto Santo and some of the Cape Verde Islands.

[7] This statement is not strictly accurate. The castle of Arguim had been built *circa* 1445. Cadamosto refers to it, and the methods of trading, in his account of his first voyage in 1455, see above, p. 17. Duarte Pacheco, *Esmeraldo*, p. 72, says that the castle was built after Prince Henry's death in 1460. This reference, and that of Barros, is probably to its completion or extension.

constableship for himself and his sons. We learnt also that at this time the islands which we now call the Ilhas do Cabo Verde were discovered by one Antonio de Nola, a Genoese nobleman, who on account of some grievances in his own country came to this Kingdom with two large ships and a *barinel*, accompanied by Bartholomeu de Nola, his brother, and Rafael de Nola, his nephew. To them the Prince gave permission to discover[1]. And sixteen days after sailing from Lisbon they reached the Island of Maio, to which they gave this name because they saw it on that day[2]; and the next day, which was the day of Sant-Iago and St. Felipe, they discovered two islands that now bear the names of these saints[3]. At the same time there went also to the discovery of those islands some servants of Prince Fernando, who discovered the others[4], which in all are ten, commonly called the Ilhas do Cabo Verde, because they lie one hundred leagues west of that cape. By the ancient geographers they were called the Fortunate Islands[5], about which we wrote at length in our *Geographia*[6]. These the King granted

[1] For da Noli and the disputed question of priority in discovery of the Cape Verde Islands, see p. xxxvi above. Barros's account differs considerably from that of Diogo Gomes, who claims to have been in da Noli's company. The islands are mentioned in a donation by Afonso V to the Infant, D. Fernando, dated 1460, but da Noli's name is not mentioned until two years later, in a similar donation: 'cinquo per Antonyo de Nolla, em vida do Ifante dom Anrrique', i.e. before 1460 (*Alguns documentos*, pp. 27, 30). The five islands mentioned in the donation of 1460 are Santa Maria, São Jacobo e Felipe, (St Thiago), Ilha dellas Mayaes, Sam Christovam and Ilha Lana.

[2] ? In that month.

[3] The names of the two saints, however, were given to one island. The day was May 1st.

[4] These remaining islands are first mentioned in the grant of 1462: their names are given as Brava, Sam Nycollao, Sam Vicente, Ilha Rasa, Bramca, Santa Lucia and Sant Atonio. They were discovered by Diogo Affonso, in the service of the Infant, D. Fernando. It has been suggested, from the evidence of the Saints' names, that the discovery was made in December 1461 and January 1462.

[5] The 'Insulae Fortunatae' of the classical geographers were the Canary Islands. Behaim on his globe of 1492 also identifies them wrongly with the Cape Verde Islands.

[6] This is another lost work of Barros—his *Geographia universalis*. From the references to it in the *Decades*, it dealt with ancient and modern geography, and the use of nautical instruments. A portion of it is said to have been in the possession of his son. No doubt its contents was similar in style to the section on the Senegal and Gambia country (Book III, ch. viii, below), and to the descriptions of the regions of India in the succeeding *Decades*.

to his brother Prince Fernando on the nineteenth of September of the year fourteen hundred and sixty two[1]. That called Sant-Iago was the first to be peopled, and by the same Prince Fernando[2], to whom the King gave the liberties which it still enjoys, by charter of the twelfth of June, fourteen hundred and sixty six. But later on, as the inhabitants used these first liberties in the trade with Guiné with more licence than the King wished, he limited them by another charter in accord with his intention when he made the first grant.

CHAPTER II

How the KING LEASED *the* TRADE *of* GUINÉ *to* FERNÃO GOMES *for* FIVE YEARS, *upon* CONDITION *that* DURING THIS TIME *he* CAUSED FIVE HUNDRED LEAGUES *of* COAST *to be* DISCOVERED; *and how for his* DISCOVERY *of the* TRADE *in* GOLD *at* MINA *he was* GIVEN *the* SURNAME *of* 'DA MINA' *with a* COAT *of* ARMS.

At this time the trade of Guiné was already very current between our men and the inhabitants of those parts, and they carried on their business in peace and friendliness, without those warlike incursions, assaults and robberies which happened at the beginning,—as could not have been otherwise with people so wild and barbarous, both in law and customs and in the use of the things of this our Europe. These people were always intractable until they grew accustomed to them. However after they learnt something of the truth through the benefits they

[1] *Alguns documentos*, p. 31.
[2] The captaincy of Ribeira Grande in St Thiago was bestowed upon da Noli, though the date of the original grant is not known. It passed to his daughter, D. Branca de Aguiar, in 1497, on her marriage to Jorge Correia (*Alguns documentos*, p. 90). This charter states that the peopling of the island had begun in 1462. Diogo Affonso was rewarded with a donation of the northern portion of St Thiago; this passed to his nephew Rodrigo in 1473. Apart from the traffic with Guinea, the prosperity of the island was increased by the introduction of orchel in 1469. For the early history of the island see Senna Barcellos, *Subsidios*, ch. i.

received, both spiritual and intellectual, and articles for their use, they became so well disposed that when ships, sailing from this Kingdom, arrived at their ports, many people came from the interior to seek our goods, which they received in exchange for human beings, who were brought here more for salvation than for slavery. As all these things were so current and usual in the parts of the coast already discovered, and as the King was very occupied with the affairs of the Kingdom and was not satisfied to cultivate this trade himself nor let it run as it was, he leased it on request in fourteen hundred and sixty nine to Fernão Gomes, a respected citizen of Lisbon, for five years, at two hundred thousand reis a year, on condition that in each of these five years he should engage to discover one hundred leagues of coast farther on, so that at the end of the lease five hundred leagues should be discovered, beginning from Serra Leoa, where Pero de Cintra[1] and Soeiro da Costa, who were the last discoverers before this lease, turned back. Later this Soeiro da Costa discovered the river which we now call 'de Soeiro'[2], between Cabo das Palmas and the 'tres pontas'[3], near the house of Axem[4], where the factory of the gold trade is. And among other terms of this contract was that all the ivory should be delivered to the King at the price of one thousand five hundred reaes per hundredweight, and the King sold it for a higher price to one Martin Eanes Boaviagem, because he had engaged to do so by a previous contract for all the ivory obtained in Guinea; and, a privilege much appreciated at that time, Fernão Gomes was allowed to buy in each of the said five years one civet-cat[5].

[1] Soeiro da Costa is not mentioned in Cadamosto's account. De Sintra passed beyond Sierra Leone to Cape Mesurado. See above, p. 83.

[2] The 'Rio de Soeiro' is the Assini River.

[3] Cape Palmas, 4° 25′ N., 7° 49′ W.; Cape Three Points 4° 42′ N., 2° 6′ W.

[4] Axim, 4° 51′ N., 2° 15′ W. A fort was built here by the Portuguese, circa 1502. This was destroyed during a native rising, and a second fort was built in 1515. Burton, *To the Gold Coast for gold* (II, p. 84), says: 'There is no better landing-place than Axim upon this part of the African coast. The surf renders it impracticable only on the few days of the worst weather'. The gold was obtained from the alluvial deposits of the Ankobr river.

[5] King, Afonso, in October 1470, declared a royal monopoly in civet, *malagueta* pepper, 'unicornio', spices, precious stones, etc. (*Alguns documentos*, p. 33). 'Unicornio' appears to be the horn of the rhinoceros, which was said by Garcia de Orta to be an antidote against poison.

This contract was made in the year fourteen hundred and sixty nine, upon condition that he was not to trade on the continent opposite the Ilhas do Cabo Verde because, as they belonged to the Infant D. Fernando, this trade was reserved for their inhabitants. Also the traffic of the Castle of Arguim was excluded, because the King had given it to his son Prince D. João as part of his revenue; however, later on the same Fernão Gomes secured this traffic of Arguim from the Prince for some years at a price of one hundred thousand reis a year. Fernão Gomes was so diligent and fortunate, that immediately in January fourteen hundred and seventy one, his agents, João de Santarem and Pero de Escobar, both knights of the King's household, discovered the traffic of gold at the place we now call the Mina, the pilots being Martin Fernandes, inhabitant of Lisbon, and Alvaro Esteves, inhabitant of Lagos. This Alvaro Esteves at that time was the most skilled man of his profession in the whole of Spain[1]. The first traffic of gold done in that land, was at a village called Sammá[2], which then had about five hundred inhabitants; and later on it was done farther down, at the place where the fortress which the King D. João ordered to be built now stands, (as we shall see in due course). This spot is called by our men 'Aldea das duas partes'[3]. But Fernão Gomes not only established this traffic of gold; his discoverers, under the terms of the contract, reached the Cabo de Santa Catharina[4] thirty seven leagues beyond Cabo de Lopo Gonçalves[5], two and a half degrees south of the Equator. During this time Fernão Gomes gained great wealth, with which he later served the King in Cepta (Ceuta) and in the capture of Alcacer, Arzila and Tangere, where the King knighted him. And in the year fourteen hundred and seventy four, which was the last of his lease, the King gave him a new coat of arms of nobility—a shield

---

[1] This description of Esteves is identical with that given by Duarte Pacheco (*Esmeraldo*, p. 118), and is one of the passages suggesting that Barros utilized Duarte Pacheco's work.

[2] Sammá, the modern Shama, in 5° 2' N., 1° 36' W., near rich auriferous deposits.

[3] 'Village of the two parts': so called because the native village was divided by a stream.

[4] Cape St Catherine, in 1° 54' S., 9° 20' E.

[5] Cape Lopez, 0° 34' S., 8° 44' E.

with crest and three heads of negroes on a field of silver, each with golden rings in ears and nose, and a collar of gold around the neck, and 'da Mina' as a surname, in memory of its discovery. And he gave him a patent to this effect on the twenty ninth of August of the said year. Four years later he appointed him to his Council, because at this time the trade of Guiné and the traffic of Mina were so profitable and assisted the economy of the Kingdom so much through the good industry of Fernão Gomes, that, both for this service and for others he had personally performed, he deserved every honour and favour that might be given to him. Also at this time one Fernão do Pó discovered the Ilha Formosa, which now bears the name of its discoverer in place of that which he had given it[1]. And the last discoverer in the life of this King Dom Afonso[2] was one de Sequeira, knight of his household, who discovered the cape which we call Cabo de Catharina, a name he then gave to it because it was discovered on the day of this Saint. And not only by order of this King, from the beginning of his government, but also by order of the same Infant D. Henrique, who, as we have seen, lived till the year fourteen hundred and sixty three[3], the conquests and discoveries continued always, including the coast whence came the first *malagueta*, which was explored by order of the Infant D. Henrique[4]; if this *malagueta* reached Italy before this discovery, it was obtained from the Moors of these parts of Guiné, who crossed through the great region of Mandinga, and the deserts of Libya, which they call Çahará, until they reached the Mediterranean sea, at a port called by them Mundi barca, or corruptly Monte da Barca[5]. As the Italians did not know where it was grown, and as it was a valuable spice, they called it *Grana paradisi*, which is the name it bears among them. The Ilha de

[1] Fernando Po.
[2] King Afonso died in 1481. Barros has already implied that Cabo de S. Catharina was discovered before the expiry of Gomes's lease in 1474.
[3] Prince Henry died in 1460.
[4] According to Duarte Pacheco (*Esmeraldo*, p. 108) the trade in *malagueta* began at the 'forest of Santa Maria' beyond Cape Mesurado. On Barros's own showing this coast cannot have been discovered during Prince Henry's lifetime.
[5] The town of Barca in Cyrenaica is now known as Merj (32° 31' N., 20° 55' E.). In the Arab period it was a stage on the caravan route from Egypt to the west. The town of Tolmeita is on the site of the former port of Barca, and it is probably to this port that Barros here refers.

WEST AFRICAN ISLANDS   111

S. Thomé, Anno bom, and do Principe[1], were also discovered by order of the King D. Afonso, and other trading-places and islands, which we do not treat in detail because we do not know when or by which captains they were discovered; however it is generally known that many more events happened and discoveries were made in the time of this King of whom we have written[2]; for example, an island, of which we still know nothing to-day, was discovered in the year fourteen hundred and thirty eight; in order that what I say may not seem strange, I shall give here evidence to the truth of which many witnesses exist. In the year fifteen hundred and twenty five a fleet of Castille of which the commander was Fr. Garcia de Loais, commendador of the Ordem de S. João, on a voyage to our Ilhas de Maluco (Molucca Islands) crossed from the coast of Guiné to the coast of Brazil; we had a *roteiro* of this voyage in which the author relates the dispute in that part of the sea between one D. Rodrigo da Cunha, an Andalucian nobleman, captain of the ship *Sant-Iago* of that fleet, and Sant-Iago Guevara, a Biscay man, captain of a *patara* also called *Sant-Iago*. They quarrelled as to who should take before the commander a Portuguese ship seized by both of them, which had come from the Ilha de S. Thomé loaded with negroes; from words the captains came to bombard shots, and at last the caravel was taken before the commander, who discussed with the pilot the possibility of taking him in his company; but he did not do so, because, there being no other person on board who

[1] Ravenstein suggests that S. Thomé and S. Antão were discovered by Sequeira on his return from Cabo de S. Catharina. In 1474 the revenues of the latter were granted to Prince João and the island was renamed Ilha do Principe. The captaincy of the Island of S. Thomé was granted to João de Paiva in 1485, and in the following year he received the donation of half the island for colonisation (*Alguns documentos*, pp. 56–7). Barros's reference to the discovery of Annobon during the reign of Afonso confirms Ravenstein's suggestion that the island was known before its 'official' discovery on January 1st, 1501, by a caravel of Fernão de Mello from S. Thomé, as recorded by V. Fernandez. Ravenstein believed that it was the island called by Behaim 'Insula Martini'. Duarte Pacheco, however, makes no reference to Annobon, and says S. Thomé and Principe were discovered by order of João II. (*Martin Behaim*, p. 28.)

[2] This statement has been used by Portuguese historians to support the theory of 'national secrecy' about the extent of the Portuguese voyages, which are held to have extended across the Atlantic some time before Columbus's discovery. (See J. Cortesão, *O sigilo nacional*.)

could take the negroes from that latitude to Portugal, this would have meant that he would be responsible for the death of many souls; therefore he kept the pilot for a day, questioning him about affairs of the sea, and then sent him away unharmed. From this pilot, according to the author of the *roteiro*, they learnt that the Portuguese were in Maluco, where they had built a fortress; and that while pursuing their voyage, two degrees beyond the Line, they had found an uninhabited island[1], called S. Mattheus, where there were two watering places, one very, and the other not so good: and on two trees there were inscriptions showing that the Portuguese had been there eighty seven years before. And it appeared to have been cultivated, for there were many fruit trees, especially sweet oranges, palm-trees, and fowl, similar to those of Spain, in the tree tops, many of which they shot down with cross-bows. He also related other things that they found there, of which I only took these as evidence of what I have said, —that at that time our men had discovered more lands than we find recorded in the writing of Gomes Eanes de Zurara. It was not unusual that they should have found this memorial cut on the trees, because at that time our men frequently did so, and some of them, in praise of Prince Henry, cut the motto of his emblem, which, as we have seen, was *Talent de bien faire*; because they thought that this inscription cut in the bark of the dragon-trees, in addition to wooden crosses, would be sufficient to establish possession of their discoveries. Later on, (as we shall see), King João the Second ordered *Padrões* [stone columns] to be erected with inscriptions setting forth when and by whom the land had been discovered. This was sufficient to establish royal possession, but now even the fortresses built there are not enough, because men have invented laws in accord with their own covetousness. And as great men spend most of their lives in works of their preference, King Afonso came to neglect the affairs of discovery, and celebrated much more those of the war in Africa by the seizure of the towns of Alcacer, Arzila and Tangier[2],

---

[1] An island 'Sam mateus' is shown on the Reinel chart of 1519 (Armee-bibliothek, Munich) approximately on the meridian of St Helena, and slightly north of the latitude of Anobon. This 'discovery' is also mentioned by Galvão.

[2] El-Qsar es Sgir was taken in 1458; Tangier and Arzila in 1471.

(as we related in our *Africa*[1]), whither he went personally: this war in Africa gave him so much satisfaction, because of the good fortune he met there, that he undertook, (if affairs of state permitted), to go personally to take the town of Fez and all its Kingdom, for which purpose he had organized an Order called the 'Ordem da Espada'. Accordingly he sent his chief Chronicler, Gomes Eanes de Zurara, to the Town of Alcacer Ceguer in Africa so that he could record as an eye-witness the feats of that war; to him he wrote a letter with his own hand in praise of the work he had performed on that mission:—not in restrained and insincere words, as Princes use, but eloquently, like an orator, and as one who was proud of his faculty[2]. Gomes Eanes, seeing the pleasure of the King in the incidents of this war, wrote the Chronicle of the taking of Ceuta, and another Chronicle of the feats of the Conde D. Pedro de Menezes, and of his son the Conde D. Duarte, relating these exploits so circumstantially and with so clear a style, that he well deserved his title. He also wrote the Chronicle of King Afonso[3], to the death of the Infante D. Pedro, and the Chronicle of King Duarte, his father. These Rui de Pina[4], who succeeded him as Chronicler, appropriated, corrected, and enlarged, especially that of King Afonso, in regard to events after the death of the Infante D. Pedro. Gomes Eanes also carried out another task in the Tombo of this Kingdom, which much clarified its contents, that is, its entry-books, rearranging in certain volumes many loose documents from the time of King Pedro to that of King João of glorious memory; for he was chief keeper of the Tombo, a post very suitable to chroniclers, as in its keeping are all the

[1] See note on p. 104, *supra*.
[2] Extracts from this letter are printed in Beazley and Prestage, *op. cit.* I, pp. xli–xlii.
[3] These remarks correctly amend his earlier statement regarding Azurara's chronicle of Afonso V. This and the chronicle of D. Duarte appeared under de Pina's name (see below).
[4] Ruy de Pina (1440–1521) was appointed 'Chronista mor' by King Emmanuel in 1497. He compiled chronicles of the earlier kings, of Duarte and Afonso V, largely based upon Fernão Lopes and Azurara, and one of João II, the result of his own researches. These are printed in the *Chronica dos seis reys primeiros*, Lisbon, 1727–9, and the *Collecão de livros ineditos da historia portugueza*, Lisbon, 1790–2. Barros's account of West African history follows de Pina's fairly closely.

papers of the Kingdom, which it is convenient should pass under its chronicler's eyes, in order that he can write more accurately and fully about the feats of the King whose officer he is. Here are kept, if he wishes and knows how to use so many documents, ordinances, parliamentary records, marriage certificates, contracts, fleets, feasts, works, gifts and grants, both in the register of the Chancery and Treasury, and the accounts of the whole Kingdom. In truth (returning to Gomes Eanes, to whom were entrusted the two offices of Chronicler and Chief Keeper of the Torre do Tombo), I do not know either how long he lived or how long he had these employments; but I know, according to what he left written by his hand, that he was not an unprofitable servant, and that, both by his style and his diligence, he deserved these posts.

BOOK THREE

Of the deeds which the *Portuguese* performed in the discovery and conquest of the seas and lands of the *Orient*: containing what was achieved in the time of the *King D. João the Second*.

CHAPTER I

How the KING, D. JOÃO, SUCCEEDING *to the* KINGDOM *on the* DEATH *of* HIS FATHER, D. AFONSO, *sent forthwith a* GREAT FLEET *to* GUINÉ *to* BUILD *the* CASTLE, *which we now call* S. JORGE DA MINA, *of which* FLEET DIOGO DE AZAMBUJA *was the* CAPTAIN-MAJOR: *and how he met* CARAMANÇA, LORD *of that place*.

As the King, D. João[1], already had in the time of his father, D. Afonso, the trade of Guiné as part of the revenue of his household, and had drawn from it gold, ivory, slaves and other things which enriched his Kingdom, and as each year new lands and peoples were discovered, his hope of the discovery of India by these seas became ever the stronger. Being a very Christian

[1] King João II, 1481–95.

prince and lord of great prudence, he ordered the building of a fortress, to be the first stone of the Oriental Church, which he wished to build in praise and glory of God for the possession he took of that which he had discovered and which remained to be discovered, through the Pope's grant[1], as we said before. And knowing that in the land through which ran the traffic of gold the negroes liked silk, woollen and linen clothes, and other domestic goods, that they displayed a clearer understanding than others of that coast, and that in the trade with our men they showed they would be easily converted, he commanded that this fortress should be erected in the place, where our men usually made the traffic of the gold. Thus, with the bait offered by the worldly goods which would always be obtainable there, they might receive those of the Faith through our doctrine, which was his principal aim. And though opinions in his Council about the building of this fortress were divergent because of the distance, and the ill-effects of the climate, the food of the country, and the labour of navigation upon those who went thither, the King considered that the possibility of getting even one soul to the Faith by baptism through the fortress, outweighed all the inconveniences. For he said that God would take care of them, since the work was to His praise, that his subjects would win profit, and that the patrimony of this Kingdom would be increased. Once the building of this fortress was decided upon, he ordered the equipping of a Fleet of ten caravels and two *urcas*,—to carry hewed stone, tiles and wood, as well as munitions and provisions for six hundred men, one hundred of whom were craftsmen and five hundred soldiers. Diogo de Azambuja[2], a man very experienced in the art of war, was Captain-major of these ships and the other Captains were Gonçalo da Fonseca, Ruy de Oliveira, João Rodrigues Gante, João Afonso,—who was afterwards killed

---

[1] Sixtus IV by the Bull of June 21st, 1481, confirmed Nicholas V's grant of exclusive rights in Guinea to the King of Portugal, and Calixtus III's grant of the spiritualities to the Order of Christ.

[2] Diogo de Azambuja (1432–1518), who had served in Spain during the Infant Pedro's reign as King of Aragon, was a trusted servant of King João and of King Manoel. The latter placed him in charge of the Castello Real built at Mogador in 1506. He captured Saffi in 1508, and was governor of the African coast south of Cape Cantin. See Prestage, E., *op. cit.*, pp. 204–5, and Cordeiro, L., *Diogo de Azambuja* (B.S.G. Lisboa, II (1892), pp. 172–249).

in Arguim, when João de Moura was Captain of that fortress—Diogo Rodrigues Inglez, Bartholomeu Dias[1], Pero de Evora, and Gomes Aires, an esquire of King Pedro of Aragon. The latter went in place of Pero de Azambuja, brother of Diogo de Azambuja, who died from plague before they sailed from Lisbon. All were noblemen and servants of the King. The Captains of the *urcas* were Pero de Cintra[2], and Fernão d'Afonso. In order to take the munitions for this fortress, they sailed some days before, and with them went Pero de Evora in a small ship, so that, if the *urcas* did not succeed in obtaining fish in the port of Bezeguiche[3], where they were to wait, this ship might do so. This business Pero de Evora with much diligence accomplished, and something more, for he made peace with Bezeguiche, Lord of that coast, from whom came the name that the port bears to-day. Diogo de Azambuja confirmed that peace on his arrival, on Christmas eve of the year fourteen hundred and eighty one, twelve days after he had sailed from Lisbon. Continuing on his course, God gave him so good a voyage—though he had some trouble with one of the *urcas*, which sprang a leak—that on the nineteenth of January in the following year he arrived at the place where he was to build the Castle, at that time called 'the village of the two parts'. There he met João Bernardes with a ship of the King trafficing in gold with Caramança[4], lord of that village. He sent him to tell Caramança that he had arrived with a great fleet, sent by the King of Portugal, his lord, on which came many noble people, for the good and honour of his person, as he would himself know later, and to ask him to be pleased to meet him next day, when he intended to go ashore. When Caramança's answer arrived, expressing delight at his coming, Diogo de Azambuja landed with all his people smartly dressed, but with hidden arms in case of need. He first took possession of a big tree on a small hill not very far from the village, a place

---

[1] The commander of the expedition which rounded the Cape of Good Hope in 1487-8.
[2] The discoverer of Sierra Leone.
[3] The district south of Cape Verde. The 'port' became a regular calling-place for Portuguese vessels on the way to the East.
[4] Caramança may be a corruption of Kwamin Ansa, i.e. King Ansa (Johnston, J. W. de G., *Hist. Geogr. of Gold Coast*, p. 12).

very suitable for the building of the fortress; on this tree he had a banner with the Royal Escutcheons hoisted, and at the foot an altar was set up, at which the first mass in those parts of Ethiopia was said. This mass was heard by our men with many tears of devotion, and thanks to God for allowing them to praise and glorify Him in the midst of those idolaters; asking Him that as He was pleased that they were the first to erect the altar of so great a sacrifice, He would give them wisdom and grace to draw those idolatrous people to His Faith, so that the church which they would found there might endure until the end of the world. As soon as this mass was said, on the day of S. Sebastião, (in whose memory this name was given to a valley in which ran a brook, where they first landed), Diogo de Azambuja drew up his men in ranks to await Caramança who had just left his village. He was seated on a high chair dressed in a jerkin of brocade, with a golden collar of precious stones, and the other captains were all dressed in silk. With the men drawn up in ranks, a long and broad way was made, up which Caramança, who also wished to display his standing, came with many people in war-like manner, with a great hub-bub of kettle-drums, trumpets, bells, and other instruments, more deafening than pleasing to the ear. Their dress was their own flesh, anointed and very shining, which made their skins still blacker, a custom which they affected as an elegancy. Their privy parts only were covered with the skins of monkeys or woven palm leaves—the chiefs' with patterned cloth, which they had from our ships. All, in general, were armed after their manner, some with spears and bucklers, others with bows and quivers of arrows; and many, in place of helmets, wore monkey skins studded with the teeth of animals. All were so hideous in these fancies, designed to convey warlike ferocity, that they provoked laughter rather than fear. Those who were considered noblemen were followed by two pages; one of whom carried a round wooden stool so that they might sit down and rest when they wished[1]; the other a war buckler. These noblemen

[1] These stools were of course emblems of chieftainship. The description of the appearance and behaviour of the king's body-guard in this passage is identical with that recorded by nineteenth-century travellers. See, for example, the accounts quoted by Ellis, A. B., *Tshi-speaking peoples of the Gold Coast*, ch. xviii.

wore rings and golden jewels on their heads and beards. Their king, Caramança, came in their midst, his legs and arms covered with golden bracelets and rings, a collar round his neck, from which hung some small bells, and in his plaited beard golden bars, which weighed down its untrimmed hairs, so that instead of being twisted it was smooth. To impress his dignity, he walked with very slow and light steps, never turning his face to either side. While he was approaching with this solemnity, Diogo de Azambuja remained very quietly on his dais, until, when Caramança was among our people, he went to meet him. Caramança took the hand of Diogo de Azambuja, and letting it go again, snapped his fingers saying, *Bere, Bere*, which means 'Peace, peace'. This snapping of the fingers is a sign among them of the greatest courtesy that can be offered. Then the king stepped aside, allowing his men to approach Diogo de Azambuja, to do the same; but the manner in which they snapped their fingers differed from that of the king: wetting their fingers in their mouths and wiping them on their chests, they cracked them from the little finger to the index finger, a kind of salute here given to princes; for they say that the fingers can carry poison, if they are not cleaned in this manner. These courtesies lasted for a good while, for Caramança brought many people with him; when they were finished and silence secured, Diogo de Azambuja began to tell him, through an interpreter, the reason for his coming,—the knowledge that the King, his Lord, had of Caramança's desire to serve him well, as he had striven to show by the rapid lading of his ships when they arrived in that part; and because these things proceeded from love, the King wished to repay them with love, which would be more advantageous than his, for it was love for the salvation of his soul, the most precious thing that man had, because it gave life, knowledge, and reason, which distinguished man from the beasts. And he who wished to know it, must first know of the Lord who made it, that is, God, the maker of sky, sun, moon, earth, and all upon it—He who made the day and night, rain, thunder, and lightning, and created the crops which nourished man. The King of Portugal, his Lord, and all Christendom, a great part of the world, recognised this God as Creator and Lord, adored Him, and believed in Him, as the One from whom they had received everything, and to whom their souls would go after death, to give account of the good and evil they had done during this life. He was a Lord so beneficent, that He took the good to Heaven, where He was, and the bad He thrust into an abyss of the earth called Hell, the dwelling of devils who tormented their souls. For Caramança to understand these things, he must be washed in holy water, which the Christians called the Baptism of the Faith; because just as the waters of the river washed the eyes, when they were blinded with dust, so that they might see better, so the baptismal water washed the eyes of the soul in order that they might perceive and understand the things of the soul. King João had sent to ask Caramança to recognize this God as his Creator and to worship Him, to promise to live and die in His Faith, and to accept baptism as a token of it; should Caramança accept and receive baptism, Diogo de Azambuja, in the name of the King his Lord promised him that henceforward the King would regard him as a friend and brother in the Faith of Christ, and would help him in all his need. As a token of that promise he had come there with these people to do him honour and to maintain his estate; and not only at that time would he receive help, but at all times while he remained in the Faith of Christ, God, and our Lord, as expounded to him. And as he had come well supplied with much rich merchandize never before seen in that land, for the safeguarding of which it was necessary to build a strong house, in addition to quarters wherein to lodge the honourable people who accompanied him, he asked him to be pleased to allow it to be built, which he hoped would be a token that the King would ordinarily send thither to traffic, from which Caramança would become powerful in his land, and lord of his neighbours, for no one would trouble him, since that same house, and the power of the King, would be there to defend him. Though Bayo, king of Sâma[1], and neighbouring princes would consider it a great honour to have this fortress built in their lands, and over and above this would render a great service to the King on that account, he was pleased that this work should be carried out in Caramança's land, because of the love and friendship which he had shown in his affairs.

[1] The modern Shama.

CHAPTER II

Of the REPLY *of* PRINCE CARAMANÇA *to the* WORDS *of* DIOGO DE AZAMBUJA: *and of the* PERMISSION *he* GAVE *for the* BUILDING *of the* FORTRESS, *which* SECURED *the* TRAFFIC *in* PEACE *to this* DAY.

Though Caramança was a savage man, yet he was of good understanding, both by nature and by his intercourse with the crews of the trading ships, and possessed a clear judgment. And as one who desired to understand what was proposed to him, he not only listened to the translation of the interpreter but watched each gesture made by Diogo de Azambuja; and whilst this continued, both he and his men were perfectly silent; no one as much as spat, so obedient and disciplined were they. At the conclusion, as though wishing to reflect upon what he had heard and to consider his answer, he fixed his eyes on the ground for a space, and then replied: he received as a mark of graciousness from the King, his Lord, the desire shewn both for the salvation of his soul and for the matters touching his honour; that he for his part certainly deserved it, for the good dispatch of the ships that had come thither to trade, they having been faithfully and honestly treated. During that time nothing had astonished him as much as the captain's arrival; on the other ships he had seen ill-dressed and ragged men only, who were content with whatever he gave them in exchange for their goods. This was the aim of their voyages to those parts; all they asked was to be dealt with immediately, since they preferred to return to their country rather than to live abroad. But with him, the captain, it was otherwise. He came with many people, and with much more gold and jewels than there were in those parts where they were found, and moreover with a new request—that he might establish a residence in that land. From this he conjectured two things: first, that the captain must be a very close relative of the King of Portugal; secondly, that a man as important as he was surely came on great affairs, such as those of God, who made the day and the night, about whom he had said so much, and whose servant his King was. But considering the nature of so important a man as the captain, and also of the gallant people who accompanied him, he perceived that men of such quality must always require things on a lavish scale; and, because the spirit of such a noble people would scarcely endure the poverty and simplicity of that savage land of Guiné, quarrels and passions might arise between them all; he asked him, therefore, to be pleased to depart, and to allow the ships to come in the future as they had in the past, so that there would always be peace and concord between them. Friends who met occasionally remained better friends than if they were neighbours, on account of the nature of the human heart—for it resembled the waves of the sea, which, breaking upon a rock barring their path, were tossed up to the sky, so that a double mischief was done, the sea being churned to fury, and the rock, its neighbour, being damaged. He did not speak thus to disobey the commands of the King of Portugal, but for the benefit of peace and the trade he desired to have with those who might come to that port: and also because, with peace between them, his people would be more ready to hear of God, whom he wished them to know; therefore, since time would reveal these inconveniences, he asked the captain to avoid them by allowing the traffic to continue as it had before. To these words and doubts, which seemed to oppose the building of the fortress, Diogo de Azambuja answered: the reason why the King his Lord had sent him with so much ceremony to that country, was his wish for peace and closer friendship with Caramança; and as a token of this he sought to build a house there in which he might keep his goods; by this His Highness showed the great trust he had in him, and in his subjects, because no one would put his goods in a place which he suspected: if there was cause for fear, he, Diogo de Azambuja, and all those who accompanied him might well be fearful, for they were trusting their lives and goods to a strange land, far from help; and though men's hearts, as Caramança had said, were free by nature, they belonged to those whose King was not so just as was the King his Lord. His subjects were obedient to his commands, fearing to disobey him more than death itself. He was neither a son nor a brother of the King, as Caramança thought, but one of the least of his subjects, and so strictly bound to perform what he had been commanded with regard to peace and concord in the work of that house, that he would prefer to lose his life rather than to disobey. These words, and the obedience they signified, so amazed Caramança that he clapped his hands, and the negroes also did this as a sign of agreement, thus interrupting the speech of Diogo de Azambuja; before he could continue, after the noise had ceased, Caramança replied that he would be pleased to permit him to build the house as he wished, warning him that peace and truth must be kept, for should our men act otherwise they would cause more harm to themselves than to him, because the land was great, and he and his men could build another abode elsewhere with a few sticks and branches, of which they had plenty. When the king had given his decision about the building of the house, but without answering the question about the Baptism to which he had been exhorted, he took leave of the captain, retiring with the same ceremony as he had come, and the captain remained with those in charge of the work, considering where the fortress should be placed. The next day the masons began to break some rocks which were close to, and overlooking the sea, where they had decided to place the foundations, but the negroes could not bear such an offence against that spirit which they worshipped as God; kindled with fury, which the devil fanned so that they should die before baptism—which some of them received later—they seized their arms, and on the impulse briskly attacked the men at work. When Diogo de Azambuja, who was with the captains unloading the munitions from the ships, saw the people running to the beach, he went in haste to their assistance; learning from the interpreter that the main cause of the tumult of the negroes was that they had not received the gifts they expected, and that they were more grieved at this delay than with the offence to their gods, he held the attention of the people as well as he could in order that there should be no bloodshed, and sent in great haste for the Factor to bring a double supply of cloths, bracelets, basins, and other things he had ordered to be presented to the king and his chiefs, as was customary; moreover, to please the negroes further he rebuked the Factor in their presence. With these gifts they were content and their fury subdued, so that they would have handed over their children, still more the rocks; so powerful is a gift, that, as it is said, Diogo de Azambuja broke both the stones, that is, the hearts of those angry Negroes, and the rocks that they defended. However during the work our men were always watchful, and careful with the negroes, lest there should be another attack. The work progressed so rapidly that in twenty days the outer wall of the castle was raised to a good height, and the tower to the first floor. On account of the special devotion of the King for the Saint, this fortress was called S. Jorge, and it was later created a town, with all the respective liberties and privileges, by a royal charter signed at Santarem on March 15 of the year fourteen hundred and eighty six. Although during this work our men continually strove to avoid quarrels with the negroes, the latter committed so many thefts and evil deeds that Diogo de Azambuja decided to burn their village. This punishment, as well as the benefits they derived, established peace firmly. When this work was finished, and trade begun in the country, Diogo de Azambuja sent back to the Kingdom the ships and the supernumeraries, with much gold that they had obtained, while he remained with the sixty men allotted to the fortress by the King's instructions. Others remained buried at the foot of the tree where the first Mass was said, which stood in the grounds of the church consecrated to São Jorge. There God is praised to-day not only by our men who visit that town, but also by those Ethiopians who, having been baptized, are included among the faithful. In this church in memory of the labours of the Infante D. Henrique, the author of this discovery, a mass is said every day for his soul by a chaplain appointed for the purpose. During the two years and seven months that Diogo de Azambuja was there, it pleased God that they did not suffer as much from disease as they had feared; and with so much prudence did he settle the prices and rules of the traffic, that to-day the greater part of his regulations are still observed, on which account the King rewarded and honoured him further on his return.

CHAPTER III

How the KINGDOM *of* BENY *was* DISCOVERED[1].

Though the christianizing of these people of Congo progressed greatly to the glory of God, through the conversion of their king, little profit accrued from what the King did in the matter of the request of the king of Beny, whose kingdom lay between that of Congo and the Castle of S. Jorge da Mina. For at the time of Diogo Cam's first return from Congo, in the year fourteen hundred and eighty six, this king of Beny also sent to solicit the King to despatch thither priests who might instruct him in the Faith. This country had already been visited in the previous year by Fernão do Po, who had discovered this coast and also an island near the land, now known by his name[2]. On account of its size he called it Ilha Formosa—but it has lost this name and bears that of its discoverer. This emissary of the king of Beny came with João Affonso d'Aveiro, who had been sent to explore this coast by the King, and who brought back the first pepper from these parts of Guinea to the Kingdom. This pepper is called by us *de rabo* (long tailed)—because the stem on which it grows comes away with it—to distinguish it from that obtained from India. The King sent some to Flanders, but it was never held in as high esteem as the Indian. As this kingdom of Beny was near the Castle of S. Jorge da Mina and as the negroes who brought gold to that market place were ready to buy slaves to carry their merchandize, the King ordered the building of a factory in a port of Beny, called Gato[3], whither there were brought for sale a great number of those slaves who were bartered very profitably at the Mina, for the merchants of gold gave twice the value obtainable for them in the Kingdom. But, as the king of Beny was very much under the influence of his

---

[1] The earlier portion of this chapter, which relates the discovery of the Congo, has been omitted.
[2] The discovery of Fernando Po is earlier dated by Barros as *circa* 1478.
[3] Gwato: 'This is the harbour of the great city of Beny, which lies nine leagues in the interior, with a good road [between them]' (Duarte Pacheco, *Esmeraldo*, p. 125).

idolatries, and sought the priests rather to make himself powerful against his neighbours with our favour than from a desire for baptism, he profited little from the ministrations of those sent thither. On this account they were recalled, and also the officers of the Factory, for the place was very unhealthy, and among the persons of note who died was this João Affonso d'Aveiro, the first to establish it. However for a considerable time afterwards, both during the life of Dom João and of Dom Manuel, this sale of slaves continued from Beny to Mina, for ordinarily the ships that left this Kingdom went to Beny to buy the slaves, and then carried them to the Mina, until this trade was altered on account of the great inconveniences which arose. A large caravel was wont to sail from the island of S. Thomé, where the slaves of the coast of Beny joined those from the kingdom of Congo, because all the vessels that sailed to those parts called there, and this caravel carried them from the island to Mina. But the King, D. João the Third[1], our Lord, who was then reigning, perceived that these pagans who were in our power passed into the hands of infidels once more, so that they lost the merit of baptism, and their souls were damned eternally; accordingly as a very Christian prince, ever more mindful of the salvation of souls than of the profits of his treasury, he ordered the cessation of this trade, although he suffered great loss by this act. And by this means there were brought into the fold of the faithful more than a thousand souls, all of whom, a year before this holy precept, were in perpetual servitude to the devil, and either remained in their original condition or became Moslems when, through the barter between the Moors and the negroes of the province of Mandinga, they came under the power of the former. For this work, done in His praise, God immediately rewarded the King: because he had placed the salvation of these heathen souls above the gaining of much gold in the slave market, another mine was found below the city of S. Jorge from which have flowed great quantities of gold down to the present time, much exceeding what he would have obtained by the sale of slaves.

[1] João III, 1521–57, the patron of Barros.

CHAPTER IV

How the KING, *on account of what he* LEARNT *from* JOÃO AFONSO D'AVEIRO *and the* AMBASSADORS *from the* KINGDOM *of* BENY, *sent* BARTHOLOMEU DIAS *and* JOÃO INFANTE *to the* DISCOVERY: *on* WHICH VOYAGE *they* DISCOVERED *the great* CABO DE BOA ESPERANÇA.

Among the many things which the King D. João learnt from the ambassador of the king of Beny, and also from João Afonso d'Aveiro, of what they had been told by the inhabitants of these regions, was that to the east of Beny at twenty moons' journey—which according to their account, and the short journeys they make, would be about two hundred and fifty of our leagues—there lived the most powerful monarch of these parts, who was called Ogané. Among the pagan chiefs of the territories of Beny he was held in as great veneration as is the Supreme Pontif with us. In accordance with a very ancient custom, the king of Beny, on ascending the throne, sends ambassadors to him with rich gifts to announce that by the decease of his predecessor he has succeeded to the kingdom of Beny, and to request confirmation. To signify his assent, the prince Ogané sends the king a staff and a headpiece of shining brass, fashioned like a Spanish helmet, in place of a crown and sceptre. He also sends a cross, likewise of brass, to be worn round the neck, a holy and religious emblem similar to that worn by the Commendadores of the Order of Saint John. Without these emblems the people do not recognize him as lawful ruler, nor can he call himself truly king. All the time this ambassador is at the court of Ogané, he never sees the prince, but only the curtains of silk behind which he sits, for he is regarded as sacred. When the ambassador is leaving, he is shown a foot below the curtains as a sign that the prince is within and agrees to the matters that he has raised; this foot they reverence as though it were a sacred relic. As a kind of reward for the hardships of such a journey the ambassador receives a small cross, similar to that sent to the king, which is thrown round his neck to signify that he is free and exempt from all servitudes, and privileged in this native country, as the Commendadores are with us. I myself knew this, but in order to be able to write it with authority, (although the King D. João in his time had also enquired well into it) when in the year fifteen hundred and forty certain ambassadors of the king of Beny came to this Kingdom, among whom was a man about seventy years of age who was wearing one of these crosses, I asked him the reason, and he gave an explanation similar to the above. And as in the time of the King, D. João, whenever India was spoken of, reference was always made to a very powerful king called Preste João of the Indies, who was reputed to be a Christian, it seemed therefore to the King that it might be possible to enter India by way of this kingdom. For he had learnt from the Abexijs (Abyssinian) priests who came to these parts of Spain[1], and also from some friars who had been from this kingdom to Jerusalem and whom he had ordered to gather information about this Prince, that his country was in the land above Egypt whence it stretched to the southern sea. Wherefore the King and his cosmographers, taking into consideration Ptolemy's general map of Africa, the *Padrões* on the coast which had been set up by his discoverers, and also the distance of two hundred and fifty leagues to the east where according to the people of Beny the country of prince Ogané lay[2], concluded that he must be Preste João, for both were hidden behind curtains of silk and held the emblem of the cross in great veneration. And it also appeared to him that if his ships continued along the coast they had discovered, they could not fail to reach the land where the promontory Praso[3] was, that is, the limit of that country. Therefore taking into consideration all these facts which increased his ardour for the design of discovering India, he determined to send immediately in the year 1486 both ships by sea

[1] Professor Prestage thinks that these were probably pilgrims to Compostella, some of whom visited Lisbon on their way.
[2] By the time of Duarte Pacheco, Ogané (Hooguanee) had been located at a little more than one hundred leagues to the east of Benin (*Esmeraldo*, p. 126). He was probably a chief in the Niger delta.
[3] The 'Prasum promont.' of Ptolemy has never been satisfactorily identified. He placed it on the east coast of Africa in approximately 15° S. Some have thought it to be C. Delgado.

and men by land, in order to get to the root of this matter which inspired so much hope in him[1].

CHAPTER VI

How a PRINCE *of* GUINÉ, *named* BEMOY, *came to* THIS KINGDOM, *because of a* WAR *in which he had* LOST HIS KINGDOM: *and how the* KING, *on account of what he had heard*, RECEIVED HIM *with* MUCH HONOUR.

When this Lucas Marcos had been despatched by the King, very contented with his gifts, there happened to arrive another Ethiopian, with whom the King was no less pleased. He was at Setubal when he learnt that a ship had arrived at Lisbon from the Castle of Arguim, on which there was a Prince from the country of Jalof, called Bemoy, accompanied by relatives and nobles of that region. The King, who, as we will see, had already heard much about him, sent word to Lisbon that he was to be well entertained, and escorted thence with honour to the Castle of the Town of Palmela. There he stayed some days, during which time he and his people were properly clothed and given riding animals, that they might go before the King; for he was treated in every respect as a sovereign Lord, accustomed to our civilization, and not as a barbarous Prince outside the law. He was similarly treated on the day of his arrival at the Court, for D. Francisco Coutinho, Count of Marialva, went to meet him, accompanied by many noblemen. On that day the King and the Queen displayed great ceremony and pageantry, each of them with their household: the King in the state room on a high dais with a canopy of rich brocade, accompanied by the Duke of Beja, D. Manuel, brother of the Queen, counts, bishops, and

[1] The remainder of the chapter deals with the voyage of Diogo Cam to the Cape. Chapter v, which describes the mission of Pero de Covilhã and Afonso de Paiva to the East, has also been omitted, as it is not immediately concerned with the affairs of West Africa. Lucas Marcos was an Abyssinian priest, sent from Rome to Lisbon in 1488. He was then despatched with letters to 'Prester John'.

other notable persons; with the Queen were Prince Afonso her son, and many noblemen of the Court and their ladies in ceremonial dress. As the account of Bemoy's speech on this occasion given by Rui de Pina, chief Chronicler of this Kingdom, in his Chronicle of the King, is as brief about the fortunes of this Prince, as it is copious in praises of the King, and in describing the astonishment of Bemoy at the sight of his state, we shall omit that eloquence, and continue with our purpose, which is to relate the causes of his exile, and what happened at his coming, because this is what concerns history. When the trade of Guiné was first established between our men and the people of this region of Jalof, which lies between these two remarkable rivers, Çanagá and Gambea, there was a very powerful king in those parts called Bór Byrão, who, although of the gentile blood of the Princes of Guinea, had become a Moor through contact with the Azenegues. Among the children he left when he died, by several wives (as was their wont), were Cybitah and Cambá, who were sons of one wife, and Byrão of another, who had been married to another man, by whom she had had this Bemoy. And because in that country when the king dies the people usually choose for their ruler the son they like best, they elected Byrão as their king. He, on assuming the government of the country, did not favour his brothers Cybitah and Cambá because they were his rivals in the kingdom on the side of his father, but showed great regard for Bemoy, his brother on his mother's side, who was not a rival for his inheritance. To Bemoy, on account of the hatred of the others, he gave the whole rule of his state, according to their custom, and completely abandoned the government and gave himself to pleasure, so that the people recognized and obeyed Bemoy alone. He, being a prudent man, seeing that our ships which carried on trade with that coast enriched the country with horses and other needful goods, and that the control of these affairs would make him more powerful, left the interior and came to the ports visited by us. He followed this prudent policy in dealing with them: if a horse died on board, he accepted its tail as sufficient proof, and paid for the horse; because he said that when it was taken on board it was already his property, and that it was not just that men who came so far to bring him what he needed should lose by it. Not only had he this manner of satisfying those he dealt with, but also in the service of King João, during whose time he lived, as a man who expected to profit by his friendship, he caused the ships, immediately they arrived in port, to be diligently dispatched; and in addition sent presents of the products of his country. Therefore the King, besides his desire to bring the Faith to all those Princes of Guinea in general, had an especial affection for Bemoy, because he was also told that Bemoy had the disposition, wisdom, and clear judgment to receive the evangelical doctrine. He always recommended the captains who went to trade in those parts to converse with Bemoy about the tenets of the Faith; sometimes he sent messengers with this request, bearing gifts and presents, and many offers to improve his state, in order to encourage him. But, either because at that time Bemoy did not deserve so much grace of God, or in order that his memory might be recorded more honourably in the chronicles of the kings of this kingdom, he did not accept baptism then, but always gave hope of it, by the contentment and pleasure with which he listened to the tenets of the Faith. This prosperity caused the death of his brother, who had entrusted to him the government of the kingdom, and his own exile; for the two brothers Cybitah and Cambá killed King Bór Birão by treason, and Cybitah, being the elder, took the title of king and began a cruel war against Bemoy. War needs men, especially if it is prolonged; so that, when Bemoy had lost some battles, his power began to wane. But relying upon his services to King João, he sent a nephew to him in a trading ship asking for aid in horses, arms, and men. To which request the King answered that if Bemoy desired help, he must receive baptism; then he would help him as his brother by law and Faith, and as a friend who had rendered him service; however, in order to comfort him and to encourage his conversion, he sent him five caparisoned horses for his use, and the Duke of Beja, D. Manuel, sent him one horse and harnesses for others. These were taken to him by Gonçalo Coelho, who was later Scrivener to the Treasury of the City of Lisbon, through whom we learnt the greater part of these affairs; with him went the messenger who had been sent by Bemoy, as well as priests to instruct him in the Faith. On this voyage of Gonçalo Coelho, some people who sailed on the trading ships ventured inland with him, seeking to sell their goods better, for on account of the war, the usual traffic did not come so regularly to the sea ports. This going to and from the encampment of Bemoy increased so much, and Bemoy bought so many horses, on account of the war, without being able to pay for them, that many people gathered there, some endeavouring to get what he owed them, and others to dispose of what they had not sold in the sea ports. Bemoy, being a cunning man, saw that the delay in giving satisfaction to Gonçalo Coelho and the others was profitable to himself, for they assisted him in the affairs of the war; he therefore kept Gonçalo Coelho there for a year, in the hope that he would secure his conversion. Gonçalo Coelho penetrated this design, and perceived also that our men were losing over the goods sold on credit to Bemoy. He therefore wrote to the King about the poor results of his work, and the loss caused by his stay. When the King received the letter of Gonçalo Coelho, he instructed him to return at once, taking leave of Bemoy without offending him, and to tell the others that they must also return under threat of heavy penalties. When Gonçalo Coelho announced that they were leaving, Bemoy was very sad, because he foresaw his downfall, as he had received great help from them in war, and also because he recognized he must pay them what he owed in order to maintain his credit. However, seeing that he could not retain Gonçalo Coelho, he paid his debts with the help of his people, and sent the same nephew, who had come with Gonçalo Coelho, to the King with one hundred good slaves taken in the war, and a thick golden bracelet, a customary credential. Among other reasons he gave to excuse himself for not receiving baptism, was that the people who followed him were at war, and that if he changed his faith it would be necessary to make them all do the same. As it is difficult to force savage people quickly to change the rites and uses in which they are born, they would more readily desert him (Bemoy) than their rites and customs. Thus the opportunity would be lost for them to receive baptism later, when he hoped with God's help to be free of his enemies. Finally it seems that God willed that matters should fall out thus, and that through these events Prince Bemoy should come to receive baptism, for he was beaten in a battle and abandoned by his men. He sought to save his life by fleeing for more than sixty leagues along the sea shore to our fortress of Arguim, where he embarked with the few who followed him, confident in the power and generosity of the King, from whom he had received so many messages and gifts, and so much honour. This confidence did not betray him; the King, remembering that he had always found Bemoy faithful in the days of his prosperity, and wishing also to secure his baptism, received him with much honour and pomp—for it is a great consolation to the miserable to be kindly received when they make their first petition. When Bemoy entered the room, the King took two or three steps from the dais, raising his cap slightly. As soon as Bemoy saw the King with his company, he threw himself on the ground at the King's feet, making as though he took the earth from under them and threw it over his head in token of humility and obedience. The King made him rise, and, returning to the dais stood leaning against a chair, while he commanded the interpreter to tell him to speak. Bemoy was a tall man, strong and good looking, forty years old, with a long and bushy beard, so that he did not seem a negro, but a prince worthy of all respect. He delivered his speech with such majesty of person, and with so many effects to arouse pity for his miserable banishment, that he was understood even before the interpreter translated his words. When he had finished his narrative, like a natural orator, placing all his hope in the generosity of the King, he was answered in a few words, but so agreeably that at once Bemoy showed another countenance, courage, and bearing; taking leave of the King, he kissed the Queen's and Prince's hands, asking them to be his mediators with the King. Thence he was taken to his lodging by the nobles who accompanied him.

CHAPTER VII

How PRINCE BEMOY RECEIVED BAPTISM, *and was* NAMED D. JOÃO BEMOY: *and of the* CEREMONIES ORDERED BY *the* KING *in his honour: and of the* CONVERSION *of* THOSE *who* ACCOMPANIED HIM.

After the day of his arrival, Bemoy was several times with the King in private, who was content both with his conversation and person; for his questions and answers revealed him to be endowed with a very clear intelligence. Among all this, that which interested the King more was what Bemoy said about some Kings and Princes of those parts, mainly of one whom he called the King of the peoples of Moses[1], whose state began beyond Timbuktu and extended towards the Orient,—a king neither Moor nor gentile, with customs in many ways like those of Christian peoples. From this the King concluded that he was the Preste João, whom he so eagerly wished to reach. The importance attached by the King to this information was very advantageous to Bemoy. The King's first care was to entrust him to the theologians, that they might expound the Faith and prepare him for baptism. He received this sacrament on the evening of the third of November, fourteen hundred and eighty nine, in the house of the Queen, his godfathers being the King, the Queen, the Prince, the Duke of Beja, a commissary of the Pope and the Bishops of Tangier and Ceuta, who also sponsored two of his principal nobles. He received the name of João. Next day, after this spiritual honour, which is eternal, he received the worldly honour of knighthood, the coat-of-arms given to him being a golden cross on a red field, with the

---

[1] The Mossi, a pagan negro people, appear to have entered the savannah region south of the great bend of the Niger, and to have established two kingdoms in the twelfth century. Their determined and successful resistance to Islam, and their use of the cross as a decorative motif no doubt contributed to the belief that they were Christian. The power of the state lay in their numerous and well-trained cavalry. They had raided Timbuktu in 1333. A résumé of their characteristics, with further references, is given in Gautier, E. F., *L'Afrique noire occidentale*, ch. v.

Portuguese escutcheons as border. In return for this honour he engaged himself to subject to the King every state he might conquer or possess; and through the Commissary he sent to the Pope his official submission, like any other Christian prince. After him twenty four of his noblemen received baptism, for which ceremony the house of the exchequer of the said town was hung with tapestries. During the celebration of the baptism of D. João Bemoy and his men, there were continual tournaments, bull-fights, farces, and great evening-parties, so pleased was the King with his conversion. D. João Bemoy wished to give displays after the manner of his country, for some of his men were very good horsemen. They rode in the presence of the King, standing, turning, sitting down and getting up again, all during one race: with one hand on the saddle-bow they jumped to the ground, with the horses at full speed, and returned to the saddle as naturally as if the horses were still. Also, at full speed, they picked up, from the saddle, stones placed on the ground, and performed many other entertaining tricks, showing themselves to be more nimble on horseback and on foot than the Alarves of Africa, who are very proud of these accomplishments. When these ceremonies came to an end, the King began to consider the question of restoring Bemoy to his state. After taking council, it was decided that the King should send with him twenty caravels equipped with people and munitions, to aid him, and also to build a fortress at the mouth of the river Çanagá. The King did not desire to erect this fortress solely to help the Prince, but resolved to do so when he learnt of the country and the course of the river, hitherto considered to be a branch of the Nile: but before we continue with the narrative of the fleet, we must speak of this river, as well as of this province of Jalof, that it may be perceived how prudent and right the King was to send so great and costly an expedition.

CHAPTER VIII

Wherein is DESCRIBED *the* LAND *that* LIES BETWEEN *the* TWO RIVERS ÇANAGÁ *and* GAMBEA, *and their* COURSES: *how* PERO VAZ BISAGUDO, *who accompanied* PRINCE D. JOÃO BEMOY *wrongfully* KILLED HIM, *on the grounds that he was* PLOTTING TREASON, *and the* GRIEF *of the* KING *at* HIS DEATH.

This land, commonly called Jalof by the natives, lies between these two remarkable rivers, Çanagá and Gambea, which, as their courses are long, receive several names from the different peoples who live on their banks. Though we call it Çanagá where it enters the sea, the Jalofs call it Dengueh; the Tucuroes, Maio; and the Caragoles, Cólle; where it runs through the Province of Bagano[1], which is the most eastern, it is called Zimbalá,—for which reason this Province is sometimes known by the same name—and in the Kingdom of Timbuktu it is called Iça[2]. Though it runs through a great extent of country, coming from the eastern springs of the lakes which are called by Ptolemy Chelonides and Nuba, and the river Gir[3], and continuing almost straight until it enters the Ocean in fifteen and a half degrees of latitude, we do not know what the other peoples call it. The reason

---

[1] A province of the empire of Mali, north of the Niger and west of Timbuktu, conquered by the Songhai in 1499.

[2] These names for the Senegal river are simply the words for 'water' in the various dialects. The Caragoles are the Soninké.

[3] Ptolemy's geography of the interior of Northern Africa is characterized by two hydrographical systems, one formed by the Chelonidas Paludes and the river Gir, with an independent Nuba Palus, the other, and more westerly, by the river Nigir, the Libyae Palus, and the Nigritis Palus. These are said to flow 'in the middle of the land'. It is improbable that either represent the Niger—Senegal or Gambia, as Barros supposed. Dr John Ball has argued that the Chelonidas Paludes are to be identified with the Kufara depression, and the Gir with a wadi to the west. Col. Tilho thinks that the former was the lowland north-east of Lake Chad. The Nigir was probably one of the wadis on the south-eastern slopes of the Atlas mountains, and some have identified it with the Wadi Gir (see *Geogr. Journ.* LXX (1927), pp. 209–11, 512).

why we generally call it Çanagá[1]—from the name of a lord of the country with whom our men at the beginning of this discovery traded—is because here they did not know any other name for it. Being a river which comes from so far, it does not bring down so much water[2], neither does the tide run so far up it as happens with the river Gambea of Cantor. It forms some islands, most of them inhabited by animals and insects on account of their wildness; in some places it is not navigable, for many rocks cross it[3], principally at a point about one hundred and fifty miles from its bar, where it is called Cólle. Here there are water-falls almost like those of the Nile. This place is called Huaba by the natives. The river runs so impetuously over them, and the rocks are cut so perpendicularly by its furious fall, that it is possible to pass dry-shod along the edge of the rocks: this however, (according to the inhabitants) can be done only when the wind blows from above, and not when it blows from below, because then the wind throws the water against the rocks, thus impeding this passage. The negroes call this place Burto, that is, 'the bow', because the water makes a column in the air until it reaches the foot. This river receives many other great tributaries, which as they come from uninhabited regions, where there are many animals, have no names among the peoples with whom we trade, much less among our men: although in the tables of our *Geographia* we place its course under graduation. Among its tributaries, there is one which comes from the south of the country called by the negroes Guinea, or Gennij, (as we shall see later); as this river flows

[1] Çanagá appears to have no connection with Zenaga, or Azanaghi. El Bekri employs the form *Senegana*, or *Sangana*. This is still preserved by the Moors in the term Isongân, applied by them to the lower Senegal valley. On the Medici portolan of 1351, a river *Senegany* is shown. The form Senegal originated in the eighteenth century (Delafosse, *op. cit.*, p. 57).

[2] Compared with the Gambia, the Senegal is an indifferent line of communication with the interior. During the wet season it is a turbulent, muddy flood, and in the dry season navigation is hindered by the exposed rocks and sandbanks. The Portuguese never made much use of it to penetrate the interior. According to Duarte Pacheco (*op. cit.*, p. 81) their ships ascended the river for sixty leagues only, to Tucuról.

[3] Duarte Pacheco speaks of a great rock, Feleuu, 250 leagues from the mouth. These are the Felu falls, 500 miles upstream. It is probable that Barros's Burto are to be identified with these, and that 'leagues' should be read for 'miles' in his text (*Esmeraldo*, p. 81).

through areas of clay, its waters are reddish, and as the waters of the Çanagá are white from that place up-stream, the Çaragoles call the junction Gufitembó[1], that is, white and red. They say that they are both competitors and contrary; because when anyone drinks water from one and then from the other, he begins to vomit, though neither of them, separately, cause this, or even after they have run together. The other river Gambea[2], of the *resgate* of Cantor[3], has not such a variety of names; the whole, as far as the *resgate* of the gold, to which our ships go, about one hundred and eighty leagues from the bar, on account of its devious course, or eighty in a straight line, is called Gambu by the natives, and Gambea by us. The greater part of it is tortuous, with many small turns, chiefly from the *resgate* downstream until it enters the sea in thirteen and a half degrees, to the south-east of the Cape which we call Verde. It brings a greater quantity of water than the Çanagá, and is much deeper, because it receives many wild tributaries with much water, which rise in the interior, called Mandinga, and their principal sources are those of the river Niguer, and the lake Libya of Ptolemy[4]. As it comes tortuously its waters break in such a manner, that they do not come with violence against our ships when these go up the river; half way to the *resgate*, it forms a small island, which our men call the 'Island of Elephants' on account of the many elephants there. Above the *resgate* of the gold there is a large rock[5]; as it obstructed the passage, this King João, of whom we speak, sent craftsmen to remove it, but this proved to be too expensive and difficult. Both these rivers Gambea and Çanagá, produce in general a great variety of fish and aquatic animals,

---

[1] The Gufitembó may be the Feleme: the red colour of the water would be due to the laterite in which the gold is found.

[2] 'Next to the Congo, it (the Gambia) is probably the safest river to enter on all the West African coast, and among all African rivers it is remarkable for a bar which can be crossed at any time of the tide....The river is tidal as far as the so-called "falls"—really shallows—of Barraconda, distant some 350 miles, and is navigable up to this point by steamers drawing 6 feet of water' (Archer, F. B., *The Gambia Colony and Protectorate*, p. 2).

[3] The *resgate* of Cantor was situated below, and near to, the shallows of Barraconda. This district is known today as Kantora.

[4] See note, p. 135, *above*.

[5] 'At Barraconda navigation is impeded during the dry season by a ledge of rocks which stretches practically across the river' (Archer, *op. cit.*, p. 2).

such as sea-horses (hippopotamus), very large lizards, which in shape and nature are like the Nile crocodiles celebrated by so many authors, and also serpents, which are small and not as monstrous as they are often painted and fabled. The animals which drink the waters of these rivers, are so numerous, and of so many varieties, that even elephants go in herds, as our cattle do here. Gazelles, pigs, panthers, and all kinds of game, of which we do not know the names, are found here in as great numbers and varieties. The land which lies between these two rivers, forms a remarkable cape, called by our men Verde, and by Ptolemy 'Arsinarium promontorium'[1]; it must be the same, although he places it in 10 2/3 degrees of latitude, and we have verified that it is in 14 1/3 degrees, according to its figure, because of the islands which are opposite it towards the West, which we therefore call the Islands of Cabo Verde, and he the 'Hesperides', and also because it lies between two remarkable rivers which he calls Darago, which is the Çanagá, and Stachires, that is, the Gambea, which in the manner of their course to the sea are much alike those we have now. However, he underestimates their course, placing their sources no great distance inland, although they come from the above-mentioned springs, to which Ptolemy gives no outlet, as we can see in his map. The land which lies between them, stretching towards the East for one hundred and sixty leagues, is generally called Jalof, and its inhabitants Jalofos, though they comprise many more of the races that Ptolemy included in the courses of the Darago and Stachio. The soil is coarse, very fertile, and heavy, especially that which is watered by these two rivers during their floods. In summer the heat of the sun makes such large clefts in this soil, that it is possible to bury a horse in one of them. To grow varieties of millet—which we call 'zaburro'—the general food of

---

[1] The identification of the features of the North-west African coast given by Ptolemy is difficult. Barros is supported by later geographers in identifying the 'Arsinarium promontorium' with Cape Verde and the two rivers with the Senegal and the Gambia on the strength of the latitudes assigned to them by Ptolemy. It is apparent, however, that the majority of his latitudes for this coast are placed too far south, and assuming that, as is probable, his 'Fortunatae insulae' are the Canary Islands, the Darago would be the Wadi Dra'a, the 'Arsinarium promontorium' Cape Juby, and the Stachires a wadi to the south (see Bunbury, E. H., *Ancient Geography*, II, ch. xxix, Pt. 2).

these peoples—they clear the silt left by the floods, then scatter the seeds without further tillage, and cover them with a thin layer of sand. They can then germinate; for if they are buried in the soil, so hard a crust forms on the surface—for the heat of the sun draws up the moisture below—that the seeds cannot spring up. The sand does not form an impediment, and the layer of soil below, saturated by the preceding flood, and by the night dews, which penetrate the sand, is sufficient for the germination and growth of the seeds. They do not grow wheat or the other seeds we use; it seems that the climate would not allow them to ripen, for the soil, especially near the Gambea, is very damp. Only in the lands inhabited by the Çaragoles, in some fields near the deserts, a small quantity of wheat, much bigger and finer than that of Spain (according to what they say) is grown. This is rather tilled with the hoe than ploughed. According to our division, this river Çanagá separates the country of the Moors from that of the Negroes, though on its borders all are half-castes, in colour and in customs, on account of the promiscuous intercourse with women of all races which is the habit of the Moors. However, regarding the quality of the land, it seems that nature placed that river between them as a boundary and a division; because that part lying to the north, principally inhabited by Moors, begins at the Western Ocean, like a band one hundred leagues and sometimes more in width, the edge of which is the river Çanagá, and continues towards the east until it is watered by the Nile. After receiving some moisture from these waters, it becomes dry and sterile again until it reaches the salty waters of the Red Sea. This desert is not so completely sterile that no inhabited places are to be found. These are the *Abeses*[1] of which Strabo writes; the rest is used as pasture by many Alarves, who traverse it in companies. They give it different names according to its qualities. The land which is covered with fine sand, without any green thing they call Çahel; that covered with herbs or bushes like poor heathland, whither they bring their cattle to pasture, they call Azagar; and that covered with gravel, like thick sand, they call Çahará. On

---

[1] 'Abeses.' There are references to the oases of Libya in Strabo, Bk XVIII, sect. 5, 24. The correct transcription of the Greek is 'auases'.

account of these conditions the inhabitants of this poor country come near this river Çanagá, and others seek out the oases of which we have spoken, which are, as it were, their orchards. By reason of this river the more inhabitable land is that which lies along its banks, where there are a number of towns; the principal, Timbuktu, lies three leagues to the north of the river[1]. Thither go many merchants from El Cairo, Tunis, Oran, Tremecem, Fez, Morocco, and other Kingdoms and dominions of the Moors, on account of the gold that is carried there from the great Province of Mandinga. They used also to go to another town called Genná[2] near this river, which in former times was more famous than Timbuktu. Either it takes its name from the Kingdom or the Kingdom from it. We call all that region from Çanagá onwards Guiné, although some negroes call it Genná, others Jannij, and others Gennij. As it is farther to the west than Timbuktu, it was usually frequented by the peoples in its neighbourhood, such as the Çaragoles, Fullos, Jalofos, Azanegues, Brabaxijs, Tigurarijs, and Luddayas, from whom, through the Castle of Arguim and all that coast, gold came into our hands. Other peoples from the interior of Mandinga come to the *resgate* of Cantor, to which our ships go by the river Gambea. And though the sands of these two remarkable rivers Çanagá and Gambea do not carry as much gold as those of our Tagus and Mondego, the opinion of men is so unreasonable that they do not appreciate so much what they have near them, as what they expect to gain through much danger and toil, such as those endure who go in search of gold to these two barbarous rivers. And since King João, of whom we are speaking, was already very well informed of these and other matters, which we have treated in our *Geography*[3], before the coming of Bemoy, by whom they were further confirmed, he thought that it would

---

[1] Barros was still of the opinion that the Senegal and Niger were identical. His position for Timbuktu in relation to the latter is approximately correct, though he was not aware that the river flowed eastwards.

[2] Jenné, on the upper Niger, in 13° 51′ N., 4° 23′ W. Barros follows Leo Africanus in deriving the name Guinea from Jenné, but it is more probably derived from Ghana, the former important trading centre considerably further north. From Barros's reference to its former greatness, he may have been confusing 'Genna' with Ghana (see Leo Africanus, III, p. 822, and Bovill, *op. cit.* p. 144).

[3] See above, p. 106, note.

be very convenient for his power, and for the good of his subjects to build a fortress on this river Çanagá, which would be a door, through which with the help of these Jalofos, whom he hoped in God would, by the agency of this Prince, D. João Bemoy, be converted to the Faith (as was the Kingdom of Congo)—he might be able to penetrate the interior of that great country and eventually to reach Preste João, whom he accounted so important for the affairs of India. Moreover, as by the Castle of Arguim, the *resgate* of Cantor, Serra Lioa, and the fortress of El Mina, a great part of the land of Guiné was bled of its gold, so this fortress on the river Çanagá would tap the gold coming to the said two markets, which lay close to its banks, and it would not fall into the hands of the Moors, who went to seek it by camel caravans across many deserts, in which sandy plains of Libya many of them were often buried on their journeys. For these and many other prudent reasons the King commanded that a fleet of twenty caravels should be equipped, as we have said, the command of which he gave to Pero Vaz da Cunha, nicknamed Bisagudo. He had with him many gallant men, soldiers, craftsmen for the building of the fortress, and some religious for the conversion of the negroes, of whom the chief was Master Alvaro, Friar of the Order of St. Domingos, his confessor, and a person very remarkable for his life and learning. But it seems that those peoples had not yet deserved from God the merit of baptism; for when Pero Vaz entered the river Çanagá with that great power, which amazed all the barbarians of that land, and while he was erecting the fortress, (which they say is built in an ill-chosen place on account of the floods of the river), he stabbed Bemoy to death on board his ship, saying that he was preparing a treason. Some maintained that Pero Vaz was deceived in this, and that what chiefly condemned D. João Bemoy to death was the fact that many people began to fall sick, because the place was very unhealthy, and that Pero Vaz was more fearful of having to remain in the fortress when completed than he was of Bemoy's treason. After the murder of the Prince, Pero Vaz returned to this Kingdom; the King was very displeased with the event, and ordered the cessation of work on the fortress of that river Çanagá. A part of these walls can still be seen today, according to our men.

CHAPTER XII[1]

Of what resulted from the sending of the GREAT FLEET *by the* KING *to* AID PRINCE D. JOÃO BEMOY, *as well in the* ALLIANCES *and* FRIENDSHIPS *that the* KING *had with* SOME LORDS *of the interior of* GUINÉ, *as in its* DISCOVERY *by* SOME MEN *whom he* SENT THERE, *until* OUR LORD TOOK HIM *from* THIS LIFE.

Although the death of Prince D. João Bemoy, as related above, changed all the purposes which the King had set before himself from Bemoy's return and from the building of the fortress, he did not abandon the trade with the rivers Çanagá and Gambea, which was carried on as usual each year. From the ships coming thence he learned that the fleet which he had sent to Çanagá had not been as unsuccessful as he thought. For, though it had not served to restore Bemoy, it had benefited the trade, and caused the interior to become better known than it had been before. The princes of those parts had been accustomed to see one or two ships only in their ports, on which were poor and ill-clothed sailors, so that they had not formed a very high opinion of the power of the King, despite all the interpreters had told them of the Kingdom. But when they saw so many ships, so many gallant people, and such warlike equipment, as went on that fleet, they were so amazed that from one to the other its fame spread through the whole of Guiné. Thus the friendship of the King came to be much more highly appreciated, and as most of them were quarrelling or fighting with one another, when they saw that the King had sent a large fleet merely to restore Bemoy, whose sole merit had been to deal expeditiously with the King's trading ships, they all began to do their best to dispatch the ships, each in his own fashion, and to send presents and promises, in their own interests and in the hope of obtaining

---

[1] Chapters IX, X and XI have been omitted; the first two deal with the Kingdom of Congo, the last with Columbus.

similar help from him should they need it, or from fear of angering him. This resulted in so much intercourse with these peoples, that the King began with more confidence to send his agents with messages to their greatest princes, and to intervene in their affairs and wars, as a known and valued friend. During this time he sent Pero de Evora, and Gonçalo Eanes to the King of Tucuról[1], and also to the King of Timbuktu, and at other times he sent, by the river Cantor[2], to Mandi Mansa[3], one of the most powerful of that part of the Province of Mandinga. On this mission went one Rodrigo Rabelo, a squire of his household, Pero Reinel, gentleman of the spurs, and João Colaço, crossbow man of the chamber, with other auxiliaries, making a total of eight persons. They took him as a present horses, beasts of burden, and mules with their harnesses and several other gifts much appreciated in that land, for they had been sent before. Of them all Pero Reinel alone escaped, being more accustomed to these parts; the others died of disease. This King was then waging war against another King of the Fullos called Temalâ[4]. On account of these and of other persons whom the King sent thither, so strong a friendship sprang up between our men and this (King) Mandi Mansa, that when in the year fifteen hundred and thirty four, in the course of my duty as Factor of the House of Guiné and the Indies, I sent Pero Fernandes to this Kingdom of Mandi Mansa, [kingdom of Mali] in the name of King João the Third, Our Lord, who reigns at present, on affairs of the *resgate* of Cantor, this king was very pleased with the royal message saying that he considered the arrival of the messenger as a good omen, because another messenger had been sent to his grandfather, who had borne the same name as he did, by another King João of Portugal. Such was the memory of the deeds of King

[1] Tekrur. The Peuls, a nomad cattle-raising people, probably of Mediterranean origin, are scattered throughout the Western Sudan. They migrated to the region west of Timbuktu, and adopted the language of the Toucouleurs, the autochthenous inhabitants of the banks of the Senegal. They later attempted to return eastwards, and the area between the Senegal and Darfur was known to Arab writers as Tekrur, from the language of the Peuls. Their capital was also known as Tekrur, and this has been identified with Podor (Delafosse, *op. cit.*, I. p. 235).
[2] The Gambia.
[3] King of Mandi, or Mali; also here referred to as king of Timbuktu.
[4] Delafosse believes that Temala was ruler of a Peul kingdom.

João among these unlettered barbarians. And not only by these, and Pero de Evora, but also by Mem Royz, a squire of his household, and Pero de Astuniga, gentleman of spurs, whom he took as a companion, the King sent messages to the King of Timbuktu, and to the above-mentioned Temalá, who was called King of the Fullos. Temalá was a fiery warrior, and at that time he rebelled in the south in a district called Futa[1] with so many followers that on reaching a river they drank it dry. The stubborn and barbarous scourge of that pagan people ravaged all it came upon. As this ferocity wrought great loss to the friends and servants of our King, especially to the King of Timbuktu, Mandi Mansa, Uli Mansa[2], he sent him sometimes messages of friendship, and others of request about the conduct of the war he was waging with these people. Also at this time he sent a letter by an Abyssinian called Lucas, who went by way of Jerusalem, to the King of the Mosés, a name famed throughout the Negroes of these parts of Guiné of which we have spoken. This prince was then at war with King Mandi Mansa. From the information that King João had of this King of the Mosés[3], and the customs of his people, he supposed him to be a subject or neighbour of Preste João, or the people of the Nobis; for he and his people had a form of Christianity, most of them bearing the names of the Apostles of Christ, in whom they believed. Also, by way of the fortress of Mina he sent to Mahomed ben Manzugul[2], grandson of Mussa, King of Songo[4], one of the most populous cities of that great Province which we commonly call Mandinga. This city lies on the same parallel[5] as Cape Palmas, about one hundred and forty leagues inland, according to its situation in the maps of our *Geography*. This Moorish king—in reply to our King's message, amazed at this novelty, (according to what

---

[1] The district of Fouta Jallon.
[2] Delafosse suggests that 'Manzugul' is probably the Portuguese attempt at rendering 'Mansa Uli'; so that the succession in the kingdom of Mali at this period was—Musa (or Mansa), Mansa Uli, and Mahomed ben Manzugul. (*See* Delafosse, *op. cit.*, p. 213.) Judging from what is said on the previous page, Mahomed ben Manzugul was probably the king to whom the emissary of King João III was sent in 1534.
[3] See above, p. 133.
[4] Delafosse states that the Fanti still call Mandinga 'Songo', so that 'King of Songo' is equivalent to king of Mali (*op. cit.*, p. 213).
[5] ? Meridian.

we have read in these letters, which are in our possession)—said that none of the four thousand four hundred and four kings from whom he descended, had received a message or had seen a messenger of a Christian king, nor had he heard of more powerful kings than these four: the King of Alimaem[1], the King of Baldac[2], the King of Cairo, and the King of Tucurol. When King João was exchanging messengers and letters with these barbarous princes, he also sent from the Castle of Arguim to the town of Huadem[3]—which lies about sixty leagues to the east of it—to establish a factory among the Moors, because there was some trade in gold there. On this business went Rodrigo Reinel[4], as Factor, Diogo Borges, as writer, and Gonçalo d'Antes as assistant. They remained there a short time only, because the country was desert, and frequented solely by those Alarves, that is, Azenegues, Ludaias, and Brabarijs, who occasionally visited the Castle of Arguim. From these they were unable to obtain the information about the interior which the King sought, for his purpose in causing these factories to be set up inland was as much to acquire knowledge of it, and to reach the lands of Preste João and the Orient, as to trade in gold. The men to whom the King entrusted these messages and discoveries, in addition to those mentioned already, were Rodrigo Rabelo and João Lourenço, his servants, and Vicente Anes and João Bispo, interpreters, whom he rewarded for their work, although they did not achieve the principal aim of their missions. And not only to his own subjects did he entrust the discovery of the interior, but also, in order that every opportunity might be taken, to foreigners such as Abyssinians, and some Alarves who came to the Castle of Arguim. He was so occupied and so eager in these affairs, that his mind was never at rest, chiefly since he saw and enjoyed many things unknown to the classical writers on this part of Africa. And as a hungry lion, from whom the game

[1] Al Yaman, the Yemen.
[2] Baghdad.
[3] Wadan, the Hoden of Cadamosto.
[4] Duarte Pacheco says of this mission: 'The late King John II had a certain Rodrigo Reinel, his squire, there (at Oadem) as factor, but these bad Azenegues treated him so cruelly that he must needs return to Portugal; indeed he only escaped from them with great difficulty and personal risk and loss' (*Esmeraldo*, p. 75).

hides fearfully in some thorny bush, which he prowls round and attacks on many sides, is wounded and hurt by the thorns, in entering and coming out, and tires of hurling himself upon the hidden prey—so the King, continually attacking on many sides this great bush of Guiné, which until today had not been entered, exhausted by this continual expenditure of his wealth, and also by the many worries arising from the affairs of the Kingdom, especially at the time of the treasons, rested somewhat from this great zeal which consumed him. Nevertheless the usual ships continued to make their voyages, until God was pleased to call him, and he was succeeded in the Kingdom by his cousin, D. Manuel, Duke of Beja, who (as we shall see) in the second year of his reign reached in the first voyage the goal towards which his predecessors had struggled for seventy-five years. It seems that divine prudence commands that, whereas some sow, others reap the fruit. Although we sometimes see this, we ought not to question the judgments of God; we must believe only that no one loses the reward of his good works, winning fame in this life and glory in the next. He Who was pleased that I, less from duty than inclination, not for reward but freely, and soliciting rather than invited, should be entrusted with recording the events of this discovery and conquest of the Orient, will allow me to go without reward,—should this work of mine deserve any—if I change or deny the merits of either. To be faithful to the achievements of King João in this discovery, we wish to note here three things this Kingdom owes to him: first, the praise of God, second, the glory and honour of the Royal Crown, and third, the increase of his patrimony. As to the praise of God, none greater could there be in His Church, than that, through the industry of this Prince, the altars of a Cathedral See, in the most remote place on earth and among people most ignorant of the name of Christ—where, we may believe, the preaching of the Apostles never reached—are today filled with offerings and sacrifices to God, in the name of Christ Jesus, our Saviour, and His Son. This Christ Jesus is believed, worshipped and confessed by a king, of barbarous blood and Catholic by Faith, with that great population of the Kingdom of Congo, which for seventy years has been in the Church of God by faith and baptism, and always increasing in faith, as we learn from the bishops, priests, theologians and ministers sent to announce the Gospel. The second thing which he left to this Kingdom, with respect to the honour and glory of his crown, are two fortresses: one in Arguim, completed by his industry, though it was begun in the life of King Afonso, his father; and the other S. Jorge da Mina, in the midst of the great region of Ethiopia. By these fortresses he established possession over what he had discovered and hoped to discover along this route, and added to the Crown of this Kingdom the lordship of Guiné which it now enjoys. As a prudent baron, and resolute prince, in order to remove doubts between his successors and the Princes of Christendom with respect to this possession, he forthwith made an agreement with King Ferdinand of Castile, delimiting what each of them might conquer, details of which can be found in the contracts and treaties made between them. As to the increase of the Royal patrimony, I do not know in this Kingdom any yoke of land, toll, tithe, excise, or any other royal tax which is more certain in each yearly return than is the revenue of the commerce of Guiné. If we know how to cultivate it, much of it will yield with little seed better crops than the crown-lands of the Kingdom, and the *leziras* of Santarem[1]. It is, besides, so peaceful a property, quiet and obedient, that—without our having to stand at the touch-hole of the bombard with lighted match in one hand, and lance in the other—it yields us gold, ivory, wax, hides, sugar, pepper, and it would produce other returns if we sought to explore it further.

[1] *Lezira* was the term applied to meadows along the Tagus which owed their fertility to the inundations of the river.

APPENDIX

NOTES ON THE NATURAL HISTORY OF CADAMOSTO'S NARRATIVE

Madeira. Cadamosto mentions several kinds of birds on the island of Madeira (p. 10 above). Quail are found to-day in the fields of maize and beans. They have certainly been on the island for a considerable period, as they are now recognized as a small dark subspecies of the migratory quail, *Coturnix coturnix confisa*. Madeira is rather far off the track of migrating birds, but quail must have first gained a footing during migration, and, finding conditions congenial, remained, eventually to become slightly differentiated. The island has always been famous for its pigeons. Besides the rock pigeons (*Columba livia*), there is also a race of wood pigeon (*Columba palumbus maderensis*) which is a resident breeding bird. The indigenous pigeon (*Columba trocaz*), a large lawrel pigeon, is restricted to Madeira, it has a near relative in the Canary Islands, but there is no pigeon in Africa or Europe remotely connected with it. It was much more common in early days, and was probably the only pigeon on Madeira at the time of Cadamosto's visit. The partridge must have been introduced into Madeira. It is the red-legged partridge (*Alectoris rufa hispanica*), and is now rare. The date of its introduction from north-west or western Spain is unknown.

There are no 'peacocks' or any birds which could be confused with them on the island. Albino peacocks are not uncommon elsewhere, but it seems unlikely that they were introduced into Madeira before Cadamosto's visit, and subsequently disappeared.

Senegal. There are great numbers of parrots in Senegal (pp. 47–8). The yellow and green parrots (*Poicephalus senegalus senegalus*) are in large flocks, and do much damage to the ground nuts. The bigger species is a large-billed bird (*Poicephalus robustus fuscicollis*). The third is a parrakeet (*Psittacula krameri krameri*), which is found across Africa to the White Nile, and might well have been exported through Alexandria. There are many species of weaver birds in Senegal, but none are green and yellow. It is clear, from his description of the nests, that Cadamosto was confusing the weaver birds with parrots.

The reference to 'Guinea hens, "galine de faraon", which are brought from the Levant,' is puzzling. The 'Poule de Faraon' is the Royal Fowl of Egypt, probably the King Reed-Hen (*Porphyrio madagascariensis*). A large purple bird, living in marsh and swamp, it is found in Senegal, but does not occur in the Levant. Cadamosto couples it with 'geese', which suggests a water bird rather than a turkey or guineafowl. Guineafowl (*Numida meleagris galeata*), however, are very numerous in Senegal, though this reference does not appear to be to them. There are no wild turkeys in Africa, though turkeys may have been imported in former times. It is possible that this passage is a confused reference to Guineafowl and King Reed-Hen.

There are three species of geese in Senegal, the commonest being the Spur-wing (*Plectropterus gambensis gambensis*), and this is probably the species referred to in the text.

It is not possible to identify with certainty the varieties of the kidney beans mentioned by Cadamosto as growing in Senegal (p. 42). The 'kidney beans,...spotted with different colours, as though painted,' agree with an unnamed species of *Canavalia* preserved in the Kew Herbarium. The other beans he describes are probably the *Canavalia gladiata* DC., which has a pod 20–30 cm. long and 3·5–4 cm. wide, with bright red seeds, and the *Canavalia ensiformis* DC., the pod of which is usually longer than *C. gladiata*, and has white seeds.

The material upon which the above notes are based I owe to Dr David Bannerman, as regards ornithology, and to the Keeper of Botany, British Museum (Natural History).

BIBLIOGRAPHY

Africa Pilot, Part I. 9th ed. Hydrographic Dept., Admiralty. London, 1930.
Alguns documentos do Archivo nacional da Torre do Tombo. Ed. by J. Ramos Coelho. Lisbon, 1892.
Almada, André Alvares d'. Tratado breve dos Rios de Guiné, 1594. Ed. by D. Köpke. Porto, 1841.
Almagià, Roberto. Intorno ad un manoscritto dei viaggi di Alvise da Mosto. (*Riv. Geogr. Italiana*, 39 (1932), pp. 169–76.)
Amat di S. Filippo, Pietro. Delle navigazioni e scoperte marittime degl' Italiani nell' Africa occidentale lungo i secoli XIII, XIV e XV. (*Bol. Soc. Geogr. Italiana, Roma*, 17 (1880), pp. 59–77, 125–45.)
Astley, Thomas. New general collection of voyages and travels. 4 vols. London, 1745–7.
Azurara, Gomes Eannes de. The Chronicle of the discovery and conquest of Guinea. Trsl. and ed. by C. R. Beazley and E. Prestage. (Hakluyt Soc., 1st ser., 95 and 100.) 1896–9.
Bannerman, D. A. The birds of tropical West Africa. Vols. 1–4. London, 1930–6.
Barros, João de. Da Asia de J. de Barros; Dos feitos, que os Portuguezes fizeram no descubrimento, e conquista dos mares, e terras do oriente. 8 vols. (*Vol.* 9, *Life, and index.*) Lisbon, 1777–8.
Bensaude, Joaquim. L'astronomie nautique au Portugal à l'époque des grandes découvertes. Bern, 1912.
Bontier, Pierre, and Jean le Verrier. The Canarian; the conquest and conversion of the Canarians in 1402 by Jean de Bethencourt. Ed. by R. H. Major. (Hakluyt Soc., 1st ser., 46.) 1872.
Book of the Knowledge of all the Kingdoms. Written by a Spanish Franciscan in the middle of the XIV century. Trsl. Sir Clements Markham. (Hakluyt Soc., 2nd ser., 29.) 1912.
Bovill, E. W. Caravans of the old Sahara: an introduction to the history of the western Sudan. London, 1933.
Burton, Sir Richard F., and V. L. Cameron. To the Gold Coast for gold. 2 vols. London, 1883.
Caddeo, R. Le navigazioni atlantiche di Alvise da Cà da Mosto, ecc. (Viaggi e scoperte, 1.) Milan, 1928.
Chevalier, Aug. Les Iles du Cap Vert: Flore de l'Archipel. (*Rev. Botanique appl.*, 15 (1935), pp. 733–1090.)
Claridge, W. Walton. History of the Gold Coast and Ashanti. 2 vols. London, 1915.
Codine, J. The life of Prince Henry of Portugal surnamed the Navigator by R. H. Major. (*Bull. Soc. Géogr., Paris*, 6me sér., 6 (1873), pp. 67–107.)

BIBLIOGRAPHY

Collecção de noticias para a historia e geografia das nações ultramarinas, publicada pela Academia Real das Sciencias. (Vol. 2, no. 1, contains Mendes Trigoso's translation of Cadamosto's narrative.) 7 vols. Lisbon, 1812–41.

Cordeiro, Luciano. Diogo d'Azambuja. (*Boll. Soc. Geogr., Lisboa*, 11 (1892), pp. 172–249.)

Cortesão, Armando Zuzarte. Subsídios para a história do descobrimento da Guiné e Cabo Verde. (*Bol. da Agéncia geral das Colónias*, no. 76.) Lisbon, 1931.

—— Cartografia e cartógrafos portugueses dos séculos XV e XVI. 2 vols. Lisbon, 1935.

Cortesão, Jaime. O designio do Infante e as explorações atlanticas até à sua morte. (*In* Hist. da Portugal, vol. 3, pt 2, ch. II.) Barcellos, 1931.

—— Do sigilo nacional sobre os descobrimentos. (*Lusitania*, fasc. 1 (1924), pp. 45–81.)

Delafosse, Maurice, see Haut-Sénégal-Niger, 1ère série.

Ellis, A. B. The Tshi-speaking peoples of the Gold Coast of West Africa. London, 1887.

Emiliani, M. Le carte nautiche dei Benincasa, cartografi anconetani. (*Boll. Soc. Geogr. Italiana*, ser. VII, 1 (1936), pp. 485–510.)

Fontoura da Costa, A. A marinharia dos descobrimentos. Lisbon, 1933 [1934].

Galvão, Antonio. The discoveries of the world. Ed. by Admiral Bethune (Hakluyt Soc., 1st ser., 30.) London, 1862.

Gautier, E. F. L'Afrique noire occidentale. (*Publn. Comité d'études hist. et scient.*, A.O.F. sér. A, no. 4.) Paris, 1935.

Goes, Damiõ de. Chronica do Principe D. Joam. Lisbon, 1724.

Haut-Sénégal-Niger. (Soudan français.) Séries d'études. 1ère sér. Le pays, les peuples, les langues, l'histoire, les civilisations. By Maurice Delafosse. 3 vols. Paris, 1912.

—— 2me sér. Géographie économique. By Jacques Meniaud. 2 vols. Paris, 1912.

Ibn Battúta. Travels in Asia and Africa, 1325–54. Trsl. H. A. R. Gibb (*Broadway Travellers*). London, 1929.

Irvine, F. R. Text-book of West African agriculture. Oxford, 1934.

Jobson, Richard. The Golden Trade, 1623. Teignmouth, 1904.

Kunstmann, Friedrich. V. Ferdinands Beschreibung der westl. Küste Afrikas (*Abhandl., histor. Kl., K. Bayer. Ak. Wissens.*, München, Bd. 8, Abth. 1 (1856).)

Labouret, Henri. Les tribus du Rameau Lobi. (*Trav. et mém., Inst. d'Ethnologie*, xv.) Paris, 1931.

La Roncière, Charles de. La découverte de l'Afrique au moyen âge; cartographes et explorateurs. V. 1. L'intérieur du continent. (*Mém. Soc. R. de Géogr. d'Egypte*, 5.) Cairo, 1924.

BIBLIOGRAPHY 153

Leo Africanus. History and description of Africa. Done into English in 1600 by John Pory. Ed. Robert Brown. 3 vols. (Hakluyt Soc., 1st ser., 92–4.) 1896.

Lopes de Lima, José Joaquim. Ensaios sobre a statistica das possessões portuguezas na Africa occidental e oriental, etc. 5 vols. Lisbon, 1844–62.

Magnaghi, Alberto. Precursori di Colombo? Il tentativo di viaggio transoceanico dei Genovesi fratelli Vivaldi nel 1291. (*Mem. R. Soc. Geogr. Ital.*, 18.) Rome, 1935.

Major, Richard Henry. Life of Prince Henry the Navigator. London, 1868.

Meniaud, Jacques, *see* Haut-Sénégal-Niger, 2 ème série.

Münzer, Jerónimo. Itinerário. Ed. by B. de Vasconcellos. (*O Instituto, Coimbra*, 83 (1932), 141.)

Mosto, Andrea da. Il portolano attribuito ad Alvise da ca' da Mosto. (*Boll. Soc. Geogr. Italiana*, Ser. III, 6 (1893), pp. 540–67.)

—— Il navigatore Alvise da Mosto e la sua famiglia. (Extr. fr. *Archivio Veneto*, 2 (1927), pp. 168–259.)

Nordenskiöld, A. E. Periplus: an essay on the early history of charts and sailing-directions. Stockholm, 1897.

Oldham, H. Yule. The discovery of the Cape Verde Islands. (*Repr. fr.* Richthofen Festschrift, 1895.)

Olivier, M. Le Sénégal. (Exposition coloniale de Marseille.) Paris, 1907.

Pacheco Pereira, Duarte. Esmeraldo de situ orbis. Trsl. by G. H. T. Kimble. (Hakluyt Soc., 2nd ser., 79.) London, 1937.

Paesi novamente retrovati. Et Novo Mondo da Aberico Vesputio Florentino intitulato. Vicenza, 1507.

Palmer, H. R. Sudanese Memoirs: translations of Arabic MSS relating to the Central and Western Sudan. 3 vols. Lagos, 1928.

Pereira, Gabriel. Diogo Gomes. As relações do descobrimento da Guiné e das Ilhas dos Açores, Madeira e Cabo Verde. Versão do Latim. (*Boll. Soc. Geogr., Lisboa*, 17 (1898–9), pp. 267–93.)

Peres, D. O caminho da India. (*In* Hist. do Portugal, ch. VII.) Barcellos, 1931.

Pina, Rui de. Chronica d'El Rey D. João II. (Collecção de livros ineditos de historia portugueza, *ed.* Jose Corrêa da Serra, vol. 2, Lisbon, 1792.)

Prestage, Edgar. The Portuguese pioneers. (The Pioneer Histories.) London, 1933.

Ramusio, G. B. Delle navigationi e viaggi. Vol. 1. Venice, 1613.

Ravenstein, E. G. Martin Behaim; his life and his globe. London, 1908.

Ruge, S. Valentin Ferdinands Beschreibung der Azoren. (27. *Jhber. Vereins f. Erdkunde zu Dresden* (1901), pp. 145–80.)

Santarem, Le vicomte de. Recherches sur la priorité de la découverte des pays situés sur la côte occidentale d'Afrique, etc. Paris, 1842.

Schmeller, J. A. Über Valentim Fernandez Alemã und seine Sammlung (*Abhandl., I. Kl., K. Bayer. Ak., München*, Bd. 4, Abth. 3 (1847), pp. 1–73.)

Seligmann, C. G. The races of Africa. (Home University Library.) London, 1930.

Senna Barcellos, C. J. de. Subsídios para a historia de Cabo Verde e Guiné. (*Mem. da Acad., Lisboa*, 2a Cl., 8 (1900).)

Severim de Faria, Manoel. Vida de João de Barros. (*In* vol. 9 of 'Da Asia'.) Lisbon, 1778.

Taylor, E. G. R. Pactolus: river of gold. (*Scottish Geogr. Magazine*, 44 (1928), pp. 129–44.)

Zurla, Placido. Il mappamondo di Fra Mauro. Venice, 1806.

—— Dei viaggi e delle scoperte africane di Alvise da Cà' da Mosto. Venice, 1815.

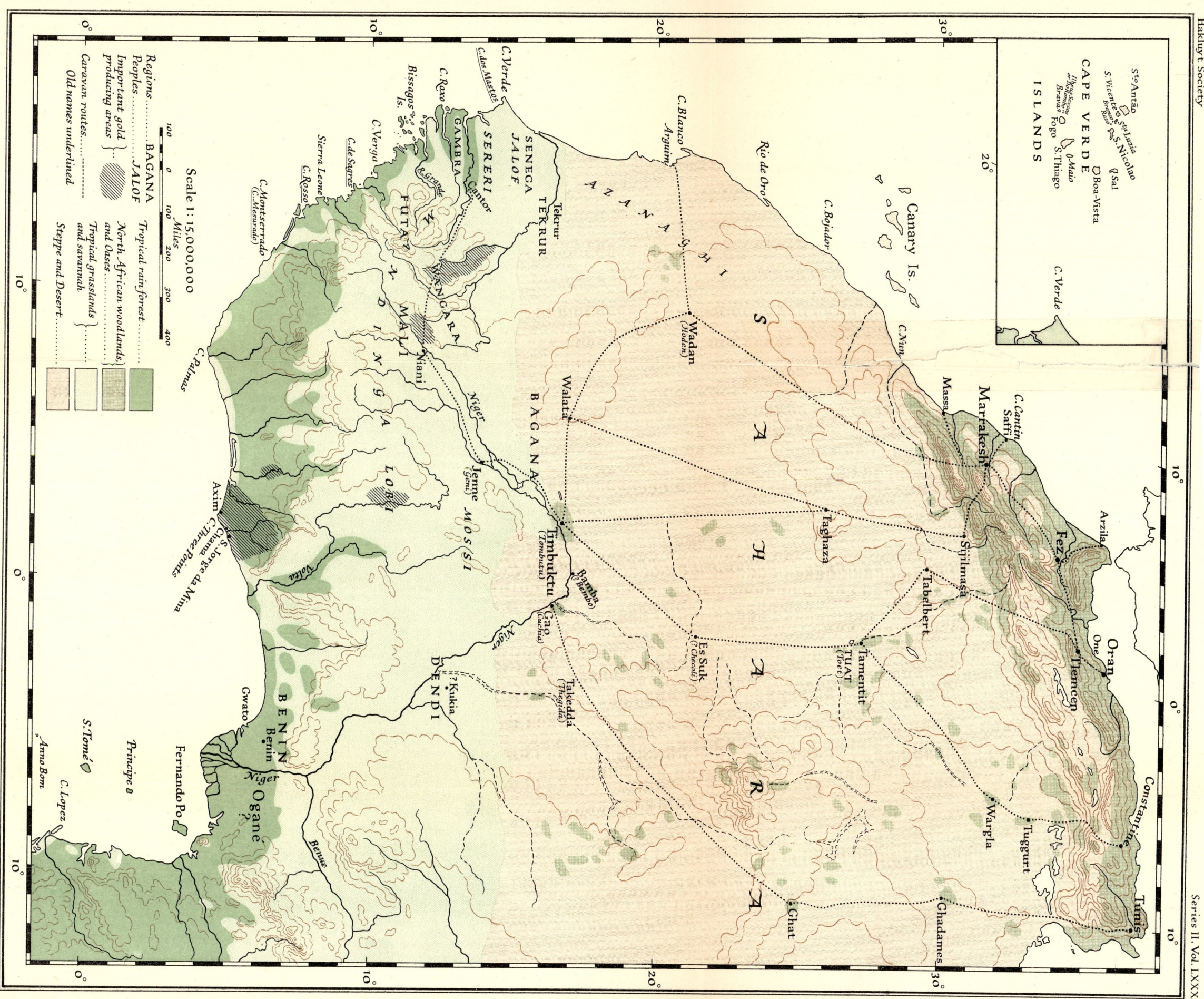

INDEX

Abyssinia ('India'), 90, 96; Abyssinian priests in Portugal, 127; *see also* Prester John
Affonso, Diogo, voyage to Cape Verde Ids., xxvii, 106 n.
Afonso, Fernão d', accompanies de Azambuja to Mina, 116
Afonso, João, one of de Azambuja's captains, 115
Afonso, King, accession, 103; reign of, 104–11; trading licence to F. Gomes, xxviii, 107
Agriculture, negro, Cadamosto on, 42; Barros on, 139
Albafur, Mountains of, 93
Alcacer dalquivi (El Qsar es Sgir), capture by King Afonso, 98
Alcuzet (? Alguechet), 96
Aldea das duas partes (Mina), 109
Alexandria, trade with W. Sudan, 89
Alvaro, Dominican friar, accompanies expedn. to Senegal R., 141
Anes, Vicente, 145
Anno Bom, Ilha, 111
Antes, Gonçalo d', at Wadan factory, 145
Arguim (Argin), xxii, 15; trade of, 17, 140; castle of, 105
Arsinarium promontorium, 138
Arzila (Arzib), 25
Assini, river (R. de Soeiro), 108
Astuniga, Pero de, mission to Timbuktu, 144
Aveiro, João Affonso d', at Beny, 124–25, 126
Axim, factory at, 108
Azagar, region of Sahara, 139
Azambuja, Diogo de, builds fortress of S. Jorge da Mina, 115 ff.
Azanaghi (Sanhaja), Tuareg tribe, xxii, 18; Cadamosto on customs of, 26; do., Malfante on, 87
Azores, peopling of, 105
Azurara, Gomes Eanes de, xix; chronicles of, 104, 112, 113

Bagano, province of, 135
Bambuk, xv

Barbazini, negro people, 54; Gomes in land of, 100
Barbazini, Rio de (the Joal), 55
Barca, Cyrenaica, 18, 110
Barros, João de, 103; *Africa* of, 113; *Asia* of, 103 ff.; *Europa* of, 104; *Geographia* of, 106, 136, 140
Batimausa (Batimansa), lord of district in Gambra, Cadamosto on, 67; Gomes on, 96
Baya, Nuño Fernandes de, with Gomes on first voyage, 91
Behaim, Martin, and D. Gomes's narrative, xxiv, xlv; on discovery of Cape Verde Ids., xxxix
Bembo (? Bamba), 87
Bemoy, Jalof prince, in Portugal, 128 ff.; death of, 141
Benincasa, Grazioso, chart of 1468, xxxiv
Beny (Benin), kingdom of, 124–25
Bernades, João, trader at Mina, 116
Beseghichi (Bezeguiche), a lord near Cape Verde, Gomes and, 99; de Azambuja and, 116
Besegue, Rio de (R. Cassini), 79
Bianca, Isola, 15
Bispo, João, 145
Bissagos, islands, 77; visited by P. de Sintra, 78
Blanco, Cape (Capo Bianco), 14
Boaviagem, Martin Eanes, contract for ivory, 108
Bofon, 87
Bonavista, Isola de, visited by Cadamosto, 64
Borgebil, King of Geloffa, 100
Borges, Diogo, at Wadan factory, 145
Bormelli, *see* Musa Mali
Bucker, a negro, 92
Budomel, Jalof king, 35–41, 99
Burbuck, King, 99
Bure, xv
Burto, waterfall, 136

Cabo das Palmas, 108
Cabo de Lopo Gonçalves, 109
Cabo de Santa Catharina, 109, 110
Cabo dos Mastos, xx, 92

INDEX

Cabo tres pontas, 108
Cadamosto, Alvise da (A. da Ca' da Mosto), voyages of, xxi–xxiv; biogr. sketch, xxx ff.; discovery of Cape Verde Ids., xxxvii; first voyage, 3–62; second voyage, 63–77
Cadamosto, Antonio, xxxi, xxxii
Çahel, region of Sahara, 139
Cairo, 25, 28, 88, 93
Cam, Diogo, 124
Camara de Lobos (Camera de Loui), 9
Cambacies (? Ghadames), 90
Canary Islands (Isole de Canaria), 11–14; birds of, 149
Cantor, trading centre, Gambia, Gomes visits, 93; Barros on, 136–37, 140
Cape Verde, 53; identified by Barros with Arsinarium promontorium, 138
Cape Verde Islands, discovery of, xxxvi ff.; Cadamosto at, 63; Gomes and da Noli at, 101, 106; granted to Prince Fernando, 107; identified by Barros with 'Hesperides', 138
Çaragoles, people, 139, 140
Caramança, King, and de Azambuja, 116 ff.
Casamanssa, Rio de (Kasamanze R.), 74
Cereculle, 95
Cerveira, Afonso, chronicle used by Azurara, 104
Checoli (? Es Suk), 87
Cintra, see Sintra
Civet, obtained in Gambia, 69; in Guinea, 108
Cochia, see Gao
Coelho, Gonçalo, emissary to Bemoy, 131
Colaço, João, mission to Mandinga, 143
Commuberta, 95
Congo, slave trade with Mina, 125
Cortese, Capo (C. Mesurado), 83
Cunha, Pero Vaz da, and proposed fort on Senegal R., 141
Cuori, Isola de, 15

Darago, identified by Barros with Senegal R., 138

De Amamento, see Tamentit
Delgado, João, 100
Dendi, 87
Dias, Bartholomeu, one of de Azambuja's captains, 116
Dias, Vincente, 6
Dragon's blood, 4, 7

Eanes, Gonçalo, mission to Tekrur, 143
Edon, see Wadan
Elephants, in Senegal, 46–47; in Gambia, 71, 137
Emin, river (? Niger), 94
Escobar, Pero de, reaches gold district at Mina, 109
Esteves, Alvaro, of Lagos, pilot, 109
Evora, Pero de, one of de Azambuja's captains, 116; mission to Tekrur, 143

Fancaso, river, 91
Farosangoli (Farisangul), King of Gambia, 67, 92, 95
Fernandes, Martin, pilot, 109
Fernandes, Pero, mission to Mali, 143
Fernandes, Valentim, and D. Gomes's narrative, xxiv, xlv
Fernando, Prince, grant of islands to, 105; servants discover some of Cape Verde Ids., 106
Ferreira, Gonçalo, 100
Ferro, 11
Fez (Fecia), 25, 90, 93
Fiumi, Rio de li, 83
Fonseca, Gonçalo da, one of de Azambuja's captains, 115
Formoso, Ilha, discovered by F. do Pó, 110, 124
Fra Mauro, map of, xxxiv
Frangazick, lord of district of Gambia, 92
Fuerteventura (Forteventura), 11
Funchal (Fonzal), 9
Futa Jallon, 144

Gambia (Gambea, Gambra), kingdom of, Cadamosto on people of, 70; 52; climate, 62; ruler of, 67; trade of, 69; Barros on, 137 ff.
Gambia, river, 56, 66, 92; attack on Cadamosto in, 58; identified by Barros with Stachires, 135

INDEX

Gao (Cochia, Cuchiam), xii, xvii, 87, 89; trade in gold, 25; do., Gomes on, 93, 95; *see also* Kukia
Garze, Isola da le, 15; Gomes at, 99
Gato (Gwato), port of Beny, 124
Gazola, land of, 86
Geloffa, land of, *see* Jalof
Geni, *see* Jenné
Genná, *see* Guiné
Ghadames (? Cambacies), 90
Ghana, 140 n.
Gnumimensa (? also Nomymans), lord of district in Gambia, 71, 97
Gold, traffic in, xiv, 22, 90, 93, 140; obtained in Gambia, 92, in Sierra Leone, 93; at S. Jorge da Mina, 123
Gomera (Giemera), 11
Gomes Aires, one of de Azambuja's captains, 116
Gomes, Diogo, voyages of, xxiv-xxv; 91–102; sights Cape Verde Ids., xxxix, 101
Gomes, FernRO, xxviii; lease of Guinea trade to, 107–10
Gonçales, Antam, 4
Gran Canaria, 11
Gufitembó, tributary of Senegal, 137
Guiné (Genná, Janni), trade of, 107; Barros on extent of, 140
Guinea hens, 48, 149

Henry, Prince, 'the navigator' (Henrique), xvii–xxi, 2–4; charter from Afonso V, 104; peoples the Azores, 105
Herrera, Diogo de, 11
Hoden, *see* Wadan
Honein (Hona, Hono), 25, 85
Horses, charms for, 49; trade in, 35
Huadem, *see* Wadan

Ibn Battuta, on trade of Tagaza, xiv
Igdem, 87
Insulae Fortunatae, 106
Ivory, from Guinea, 108

Jacob, 'an Indian', 96
Jalof (Geloffa, Jolof, Wolof, Zilofi), negro people and region, Cadamosto on, 29; customs and religion, 31; warfare, 33; methods of agriculture, 42; markets of, 48;

Portuguese and prince of, 128 ff.; Barros on, 135; do., on agriculture of, 139
Jenné (Geni, Genna), xv, 87, 140
João I, King, 2
João II, King, and trade of Arguim, 109; *Padrões* erected by order of, 112; accession and reign of, 114 ff.; achievements of, 146–47
João III, King, and slave trade, 125

Kukia (Geugeu, Quioquun), xvii; *see also* Gao

Lanzarote, 11
La Palma, 11
Liedo, Capo, 81
Lobi, xvi, 23 n.
Los, Isles do (or Idoles), 80
Lourenço, João, 145
Lucas Marcos, Abyssinian priest, visits Lisbon, 128; sent to king of the Mosés, 144

Machico (Moncricho), 9
Madeira, 8–10; birds of, 149
Mahomed ben Manzugul, King of Songo (? Mali), 144
Maio, island, 106
Malagueta, coast, 110
Malagueta, obtained by D. Gomes, 91
Malfante, Antoine, letter from Tuat, 85–90
Mali (Melli), xii, xv, 87; trade in gold, 21; Portuguese mission to, 143
Mandi Mansa (King of Mali?), Portuguese mission to, 143
Mandinga, province, gold from, 140; Portuguese mission to, 143
Massa (Amessa, Messa), 25, 90
Melli, *see* Mali
Mendes, Soeiro, Constable of Arguim, 106
Mesurado, Capo, 83
Moncricho, *see* Machico
Montagna Liona (Sierra Leone), P. de Sintra at, 81
Monte, Capo del (C. Mount), 83
Morocco, 25
Mosés, people of, *see* Mossi
Mossi, negro people of Niger, 133; message sent to king of, 144

Moura, João de, Captain of Arguim, 116
Muhammadanism, among Azanagi, 19, 87; among Jalof, 31, 40; in Gambra, 70, 97
Musa Mali (or Mansa Musa), xiii, xv; confused by Gomes with ruler of Gao, 95; *see* Mali

Niger, confused with Nile, 88; with Senegal, 28; perhaps the Emin of Gomes, 94
Nile, 28; confused with Senegal, 88
Noli, Antonio da, xxvi; voyage to Cape Verde Ids., 100, 106; captaincy of Ribeira Grande given to, 102, 107
Nomymans, lord in Gambia region, 97, 100; *see* Gnumimansa

Oden, *see* Wadan
Ogané, West African King, 126
Oliveira, Ruy de, one of de Azambuja's captains, 115
Oran, 25

Palm wine, 42
Palme, Fiume de le (Sulima R.), 83
Parrots, 47, 149
Pepper, *de rabo,* 124; *see also* Malagueta
Perestrello, Bartholomeu, 7
Philistines, *see* Azanaghi
Pina, Rui de, chronicles of, 113
Pó, Fernão do, discovers island, 110, 124
Pole Star, Cadamosto on, 61; Gomes on observation of, 101
Porto Santo, 7
Prado, de, 101, 102
Prasum, promontory, 127
Prester John (Preste João), xix, 90; Ogané thought to be, 127; King of Mossi thought to be, 133, 144
Principe, Ilha do (I. de S. Antão), 111
Ptolemy, map of Africa, 127; geography of Northern Africa, 135, 137–38

Quadrant, Gomes on use of, 101
Quioquun, *see* Gao

Rabelo, Rodrigo, sent on mission to Mandinga, 143, 145
Ramusio, Giovanni, and Cadamosto's voyages, xliii–xliv
Reinel, Pero, sent on mission to Mandinga, 143
Reinel, Rodrigo, factor at Wadan, 145
Ribeiro, João Gonçalves, on Gomes's first voyage, 91
Rio Grande (R. Jeba), 75; people of, 76
Rodrigues Gante, João, one of de Azambuja's captains, 115
Rodrigues Inglez, Diogo, one of de Azambuja's captains, 116
Rossa, Isola, 82
Rosso, Capo (Roxo), 75
Rosso, Capo, 82
Rosso, Fiume (Cockboro R.), 81
Royz, Mem, sent on mission to Timbuktu, 144

Saffi (Azafi, Zaffi), 25, 90
Sagoto, 87
Sagres, Capo, 79
Sahara (Çahará, Sarra), 15, 86, 110; Barros' account of, 139
Sambegeny, a lord in W. Sudan, 94
Sammá (Shama), traffic in gold at, 109; king of, 119
Sancta Anna, Capo de, 82
Sancta Anna, Rio de (Cacheo R.), 75
Sancta Maria, Boscho de, 83
Sancta Maria, Rio de (Bagru R.), 82
Sancto Dominico, Rio de (Mansoa R.), 75, 91
San Iacomo, Isola de, visited by Cadamosto, 65; *see also* Santiago
Santa Cruz, 9
Santarem, João de, reaches gold district at Mina, 109
Santiago, Cape Verde Ids., Gomes at, 101; voyage of da Noli to, 102, 106; *see also* San Iacomo
San Vincenzo, Rio de (Forikaria R.), 80
São Antão, Ilha de, *see* Principe
S. Jorge da Mina, xxviii; reached by J. de Santarem and P. de Escobar, 109; fortress built by D. de Azambuja, 114 ff.; trade at, 124
S. Mattheus, Ilha de, 112

INDEX

S. Thomé, Ilha de, 111; and slave trade, 125
Semanagu, a lord in W. Sudan, 94
Senegal, kingdom of, Cadamosto on, 29; products of, 42; fauna of, 46–48; Barros on, 135 ff.
Senegal, river (Rio de Senega, Çanagá), 18, 27; 135; identified by Barros with Darago, 138; attempt to build fortress on, 141
Sequeira, de, discovers C. de S. Catharina, 110
Sereri, negro people, 54
Serra Geley (Mount Gelu), 93
Sierra Leone, 141; P. de Sintra at, 81; gold in mts of, 93
Sijilmasa, xiii
Silent trade, xvi, 22
Sintra (Sinzia), Pero de, xxvii; voyage beyond Sierra Leone, 78–84, 108; accompanies de Azambuja to Mina, 116
Slave trade, at Beny and Mina, 124–25
Soeiro da Costa, voyage to Sierra Leone, 108; discovers Assini river, 108
Somanda, 95
Songo, kingdom (? Mali), Portuguese mission to, 144
Soto de Cassa, Abbot of, 100
Southern Cross, 61
Stachires, identified by Barros with Gambia R., 138
Strabo, quoted on oases, 139
Sudan, Western, Malfante on, 88; see Gold

Taghaza (Tagaza), position, xiii; trade of, Ibn Battuta on, xiv; Cadamosto on, 21
Takedda, 87
Tambucutu, see Timbuktu
Tamentit, 86
Teixeira, Tristão Vaz, 8

Tekrur (Tecuról), 34; Portuguese mission to, 143
Temalâ, a king of the Fullos, 143–44
Tenerife, 11
Thambet, see Timbuktu
Thegida, see Takedda
Thora, 87
Tides, in mouth of Rio Grande, 76
Timbuktu (Tambucutu, Tanbutu, Tenbuch, Thambet), xii, 87, 93; situation of, 140; trade of, xv, 21; Portuguese mission to king of, 143
Tlemcen (Trimicen), 90
Trimicen, see Tlemcen
Tripoli, 90
Tristão, Nuno, voyages of, 3 n., 15 n.; death of, xx, xxv
Tuareg, see Azanaghi
Tuat (Tueto), 25; organisation and trade, 85
Tunis, 25, 90, 93

Uli Mansa, King of Timbuktu, Portuguese mission to, 144
Usodimare, Antoniotto, xxiii; joins Cadamosto, 52; accompanies him on second voyage, 62

Velho, Gonçalo, and the Azores, 105
Verde, Rio (Mellakori R.), 81
Verga, Capo de, 79

Wadan (Edon, Hoden, Huadem, Oden), 87; trade of, 16, 25; Portuguese factory at, 145
Wangara, gold of, xv, 22
Weaver birds, 48, 149

Zaffi, see Saffi
Zamor (Azamor), 90
Zarco, João Gonçalves, 9
Zaya, port, 100
Zilofi, see Jalof
Zuchalin, King of Senegal, 29
Zurara, see Azurara